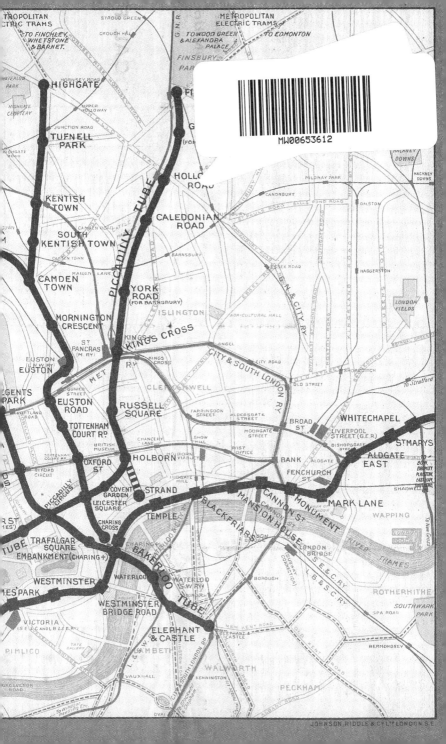

RUTH BELVILLE

RUTH BELVILLE

The Greenwich Time Lady

David Rooney

For Pat, Terry and Pete

First published by the National Maritime Museum, Greenwich, London SE10 9NF
www.nmm.ac.uk/publishing

© 2008, National Maritime Museum, Greenwich, London

ISBN 978-0-948065-97-2

A CIP catalogue record for this book is available from the British Library.

Publisher: Rachel Giles
Project manager: Emily Winter
Copyeditor: Bernard Dod
Indexer: Enid Zafran
Designer: Lana Le
Typesetter: Lisa Cutmore
Printed and bound in England

Mixed Sources

Product group from well-managed
forests, controlled sources and
recycled wood or fiber
www.fsc.org Cert no. TT-COC-002303
© 1996 Forest Stewardship Council

CONTENTS

INTRODUCTION

Ruth Belville in the Evening News, *1929.*

WHAT TIME IS IT?

It's a simple question and this book looks at some of the ways we have tried to answer it over the last couple of hundred years. But this is not just a history book, since the question is one we still ask today. Of course, our watches can now keep time very well but they need to be set in the first place. How do we know what time to set them to?

There are many ways to find out. We can listen to BBC radio where, on the hour, just before the news bulletins, we hear the six short 'pips', telling us precisely when the new hour begins. Or in London we might hear the live sound of Big Ben, the hour-bell of the great Westminster clock, tolling its ponderous temporal message as it has done since 1859, a message also broadcast daily across the world through the BBC transmitters. But we might want to know the time *now*, without waiting for the radio, and for this we can pick up the telephone and call the speaking clock. Every ten seconds, its recorded voice tells us what the time will be *precisely*: at the third stroke. There are clocks and watches that set themselves right, without our intervention. They listen to their own special radio broadcasts, pick up the rhythm, correct their micro-circuits and set their own beat. Our computer clocks also do this down a live internet connection, if we know how to set them up. We're surrounded by signals, and we have timekeepers embedded in our microwave ovens, central-heating thermostats, mobile phones and television sets. We hardly need to think about the time of day. It's always there, marking the rhythm of our daily lives.

Although the technology of timekeeping is important, this book is really more about people. The people who design and operate timekeeping systems but also those who use them – you and me. Neither the BBC six pips nor the speaking clock have been around forever and computers are mere newcomers to the everyday scene. So before the advent of these modern time sig-nallers, how did we know the time – and why did it matter?

The focus for our story is one family, the Belvilles, who from the 1830s to the 1940s provided a weekly dose of Greenwich time for those prepared to pay for the privilege. It was a small family comprising the father, John Henry Belville (often called simply John Henry), his third wife, Maria, and their only child, Ruth Belville, described by a historian who met her in the 1930s as the 'Greenwich time lady'. John Belville was an astronomer and mete-orologist at the Royal Observatory, Greenwich, who set up a time-distribution service based on sending a corrected pocket watch round to London subscribers. On John's death in 1856, the busi-ness was taken up by Maria, who carried Greenwich time to the same clientele until 1892. Then, old and infirm, she was forced to retire and Ruth took over, keeping up the weekly visits for the best part of fifty years.

From start to finish of the Belvilles' service the timekeeper used was a single pocket watch, a silver-cased chronometer of 1794 by top maker John Arnold, although understudy watches joined the business as time went on. 'Arnold', as it was affectionately known by Ruth, has survived the whole family and is on display in the Clockmakers' Company museum in London's Guildhall. Arnold's steely, mechanical heart kept beating throughout the Belvilles' lives, remorselessly counting out thousands of millions of seconds, day in, day out, as the mute carrier of Greenwich time from one place to another. Its source of accuracy – the Royal Observatory at Greenwich – has seen great changes since then: it is now a thriving public attraction, welcoming a million visitors each year as part of the National Maritime Museum. There was also great change in the Belvilles' lives, but reassuring continuity too. Both the Observatory and the Clockmakers' Company feature largely in their story and are still going strong today.

The heart of the book, and the focus for two chapters, is an episode that took place in 1908. This revolved around nothing more earth-shattering than a lecture and a few newspaper reports, but the ramifications were wide and the context highly significant. It also helps to answer another question: why did a woman continue to carry a pocket watch around London, selling time in that seemingly old-fashioned way, into the 1940s? In 1908, the lecturer and reporters already thought Ruth Belville an anachronism. She proved them wrong, but the point they made was interesting.

So this book also looks at why old habits die hard, taking a sideways glance at a period when many things changed but many stayed the same. It is about the life and death of men and women, their hopes and fears, ambition and pride, success and failure, joys and sorrows, and much in between. Its scale is as big as the world and as small as the cottage in Maidenhead where Ruth lived in 1908. Ultimately, the story is told by the people it covers, using original documents, newspaper and other reports, maps and photographs. Much of it can still be retraced through modern London, in particular buildings and transport routes, which are remarkably enduring despite the combined efforts of First World War Zeppelins, Hitler's Luftwaffe and later town planners and developers. Readers may wish to arm themselves with maps of London – old and new – to keep track of the streets and routes described, since this is a story about people constantly on the move. It has also been written for a wide readership and, in particular, for those with no detailed knowledge of timekeeping history. In the interests of readability there are no footnotes, but every source is listed at the end, chapter by chapter.

Our account begins with the birth of John Belville in 1795, and ends with Ruth Belville's death almost 150 years later. At least six of Britain's fifteen Astronomers Royal make an appearance, along with businessmen and property developers, millionaires and murderers, the Greenwich time lady and the girl with the golden voice. There are scientists, telephonists, terrorists and horologists, poets and paupers, bombers and bell-ringers; also Albert Einstein and Dash the dog, and if the story seems at times sensational, then perhaps fact really is stranger than fiction. Songs are in the air, messages on the wires, stars in the sky and blood on the streets. There are long lives and a lot of change, but all the way through people are asking a simple question: what time is it?

Chapter 1

JOHN BELVILLE AND THE ROYAL OBSERVATORY

{ 1795-1856 }

PUNCH'S FANCY PORTRAITS.—No. 134.

SIR GEORGE B. AIRY, K.C.B., F.R.S.

THE ASTRONOMER-ROYAL WHO DESERVED THE GRATITUDE OF HIS COUNTRY
FOR HAVING "CORRECTED THE ATMOSPHERIC CHROMATIC DISPERSION."

Astronomer Royal George Airy in Punch, *1883.*

ROYAL OBSERVATORY, GREENWICH, APRIL 1811. Two men had recently started working at the Royal Observatory. One was John Pond, who was taking up the position of Astronomer Royal that had been made vacant by the recent death of Nevil Maskelyne. The other was a fifteen-year-old named John Henry Belville. Belville was not yet on the official staff of the Observatory. That would come five years later, but in 1811 he was here because John Pond was his guardian. The lad settled quickly in his on-site lodgings, a tiny bedroom above one of the telescope rooms looking onto the Observatory's front courtyard with its breathtaking views over London's river Thames. He may not have started officially, but immediately threw himself into the scientific life of the Observatory, working day and night at his labours: and he stayed for forty-five years.

The early history of John Henry Belville is hard to pin down, but this is what we believe. He was almost certainly born in 1795, probably on 21 July. His pregnant mother had recently escaped the terrors

13

of revolutionary France and fled to England; and it is just possible that John may have been born in France before his mother left. As for the father, we don't know. In England, mother and child settled in the West Country, where John Pond happened to be living. One of the Observatory's twentieth-century archivists, Philip Laurie, said that Pond was actually the father, but while he may have been travelling in France at about the right time, there is no real evidence for this. Pond took the boy under his wing, and when he was appointed Astronomer Royal brought Belville to Greenwich. His name had to change, though. His daughter, Ruth, later explained what happened: 'for many years Mr John Pond thought it advisable for my father to drop his surname Belville and to be known as Mr Henry, as the horrors of the French Revolution combined with the wars in France would have prevented my father from obtaining a post under Government if of French origin.' It was not until much later in life (long after John Pond had died) that Belville relaxed, publishing two books about scientific instruments under his full name.

In 1819, still living at the Observatory, he married his first wife, Sarah Dixon, and in 1822 the couple moved out of their little room to set up home in Park Row, a street running up from the river Thames, past what is now the east wing of the National Maritime Museum, to the northern boundary of Greenwich Park. Almost three years later, in July 1825, they moved again, just round the corner to Park Terrace (now Park Vista). The following year, Sarah Belville died, having recently given birth to their fifth child. In 1827, John married Apollonia Slaney, who soon gave birth to their only daughter, Cecilia. Then, on 22 February 1833, the couple moved to another Greenwich property, on Blackheath Road, the western part of the main thoroughfare between London and Dover, rising steeply to the heights of Blackheath at the southern edge of Greenwich Park.

That year, John Pond installed on the Observatory roof the world's first Greenwich time signal. The Observatory sits on top of a hill and can be seen for miles around, a perfect location for a visual time-check. At one o'clock every afternoon, a five-foot black time ball dropped from the top of a 15-foot deal mast fixed to the north-east turret of Flamsteed House, signalling the precise time at Greenwich to mariners in the Thames and docks below. The time ball is still there and it still works, although it has been red since 1919 when the leather-covered ball was replaced by an aluminium one. John Belville was the man originally in charge of its operation and it was a huge success. A few years later, the *Illustrated London News* carried a full-page illustrated article describing the workings of the new signal: 'from the beauty of the instruments, the exactitude of the observations, and the high scientific ability of the officers engaged, the once difficult problem of finding the precise instant when one o'clock touches the world's history, is no longer in a matter of doubt or difficulty.' It started a trend. Brother and sister time balls sprang up around Britain and overseas, many triggered remotely from Greenwich. But John Pond was not to see his scheme flourish. In increasingly poor health and often away from the Observatory on sick leave, he retired in 1835 and, on 7 September the following year, he died. He was buried in the old churchyard in Lee, a short walk from the Observatory, in the Admiralty tomb containing the remains of Edmond Halley, second Astronomer Royal and the man famous for predicting the return of the comet that henceforward bore his name.

The assistants working at the Observatory were worried. Their old chief had not been much of a disciplinarian, but they had heard that his successor, George Airy, Director of the Cambridge Observatory, was a very different man altogether. They were right: Airy arrived in 1835 and got straight to work by sacking the First Assistant, Thomas Taylor, trying and failing to sack the Fourth Assistant, William Richardson, and watching cheerily when the Third Assistant, Frederick Simms, resigned because he was not promoted to replace Taylor. It's a safe bet that John Belville, the Second Assistant, was keeping his head down trying to keep out of the firing line, but luckily for him Airy liked hard workers and Belville lived to fight another day. But the newly arrived Astronomer Royal was not so happy with Pond's professional priorities. What should have been a scientific institution – a place for carrying out vital astronomical research for the advancement of knowledge and the nation – was instead, in his words, merely 'a place for managing Government chronometers'. And the man in charge of the chronometer work was John Belville.

Airy had been looking through the letter-books kept by Thomas Taylor. 'I found that out of 840 letters, 820 related to Government chronometers only.' The fault, Airy felt, lay in the Admiralty departments he had to deal with – or rather, as he put it, 'the inferior departments of the Admiralty, especially in the Hydrographic Office'. This department was responsible for marine navigation and charts, and was run by Captain Francis Beaufort, famous today for the wind scale that bears his name, but at the time he was responsible for sending hundreds of marine chronometers to the Greenwich Observatory every year for tests. Airy found it difficult to conceal his annoyance that Beaufort's chronometers were creating such a burden on the Observatory. 'I had some correspondence with Captain Beaufort, but we could not agree, and the matter was referred to the Admiralty.' Beaufort was furious, but Airy went further. On 4 June 1836, the Board of Visitors (the Observatory's governing body) was at Greenwich for the annual 'visitation day', and Airy was delivering his report. 'I did not suppress the expression of my feelings about chronometer business,' he later recalled in his autobiography. Standing before the visitors, he tersely criticised the effect the chronometers were having on the astronomical work of the Observatory, concluding 'that the persons of this establishment are astronomical observers and calculators, not clerks; that the Observatory is an astronomical institution, not a storehouse; and that any regulation which makes the account-keeping and store-keeping department predominant over the astronomical is an unjustifiable and injurious diversion of its powers.'

Beaufort, the prime target of Airy's rant, was one of the Official Visitors present in the room, and he was not happy. He tried to suppress the publication of Airy's report, but the astronomer eventually had his way, changing the rules surrounding the chronometer work and reducing the frequency of interruptions to his own. 'Chronometer-makers', it was announced, 'are admitted to the Royal Observatory, for the purpose of inspecting and removing the chronometers, and for other business, on Mondays only.' These visits were not just to drop off and pick up chronometers. The craftsmen, who worked in central London, out of sight of Observatory Hill, had long visited Greenwich for something less tangible – the time.

Anybody making a precision watch or chronometer needed to know the right time. To set their watches up correctly, to make proper adjustments so that their new little machines would beat time correctly, these men needed a 'time standard', a master clock in their workshop which could be used, day after day, as a time-check for the devices under construction. These time standards were similar in principle to domestic grandfather clocks but much more accurate.

They were known as 'regulators' and they were the same breed of clock that the astronomers themselves used. Without a good regulator, watches and chronometers could not be as accurate or as stable as was possible, and for the users of chronometers, in particular, that could mean the difference between life and death.

A chronometer is a very accurate portable timekeeper designed for one specific purpose: to keep time at sea, for months on end, so that the navigating officer can find the ship's position east or west of a fixed point. This is called longitude and it was a real devil to find before John Harrison completed the first practical marine timekeeper in 1759. What does time have to do with distance? The concept is simple. As we all know, the time of day at different places on Earth differs at any instant. The time in London is different from the time in New York, for instance, which is why we change our watches when we travel. The Earth spins on its axis once in 24 hours, so 'local noon' appears to sweep its way round the globe once every day.

A navigating officer wants to know how far east or west he is from a fixed point on Earth – Greenwich, for example. He's in the Atlantic, perhaps, and can see no land to tell him how far he is from Greenwich, but he can find his local time very easily. This is the time according to the Sun wherever he is. When the Sun is at its highest point in the sky, that, by definition, is noon. This 'local time' is easy to find with a simple angle-measuring device such as a sextant. If the navigating officer also knows the local time at the *same instant* in his reference place – Greenwich, in this example – then the *difference* between the two times is equivalent to the east–west *distance* between his ship and Greenwich, because one hour of time-difference equates to 15 degrees longitude-difference or, put the other way round, one degree of east–west distance is equivalent to four minutes' difference in time. All the navigator needs is a clock or watch, set to Greenwich time before setting sail, that will keep going and stay accurate in the rough and tumble of a sea voyage, with its extremes of temperature and its damp, salty air. Easy? In principle, yes, but in practice, not at all.

Great scientists, including Robert Hooke and Christiaan Huygens in the seventeenth century, had tried and failed to design such a time-keeper, and the idea was later damned entirely by Isaac Newton. But John Harrison, a self-trained clockmaker from Lincolnshire, spent his life proving them wrong. By the early nineteenth century the portable and accurate timekeeper he developed was becoming a standard accessory on ships, after other makers had worked out how to manu-facture the machine cheaply and in great numbers. By then it was called the 'chronometer' – time-measurer – a name given to a pocket-sized timepiece by John Arnold that performed so well on trial at the Royal Observatory that Arnold and his supporters said it deserved a new name.

So we got the chronometer, the machine which enabled captains to find their position accurately, and which could prevent them run-ning onto rocks because it had been miscalculated. Of course the base-line from which time was measured could be anywhere that had an observatory to do it, and until Greenwich time became the standard for the world after 1884 there were lots of different such baselines. French time was based on the Paris Observatory's meridian, Russian on that of St Petersburg and so on. But the London chronometer-makers needed Greenwich time when making the machines, to adjust them correctly and get them beating exact seconds. London was also the world's largest port and Britain the world leader in maritime trade. Ships' captains also needed Greenwich time, using it to set their chronometers just before leaving London on their long voyages. So it is easy to see why in the late eighteenth century and into the nineteenth, with this twofold demand, more and more people needed to know the time kept by the astronomers atop Observatory Hill in Greenwich – lives and world trade depended on it.

But there was no telephone in the early nineteenth century, no radio time signal. The electric telegraph was yet to be invented. How could you find the time at Greenwich? If you could see the Royal Observatory, as captains of ships tied up in the East India Docks could, then, from 1833, you could set your chronometer from the time ball. But before then, and for everybody who could not see the Observatory, there was no easy practical source of Greenwich time.

There is also the obvious question of how the Observatory astronomers knew the time so accurately in the first place. What did they have that ordinary people didn't? For most of human history we have measured time by the rotation of the Earth, once per day. This rotation makes it look as if the Sun rises and sets, but it is us moving, not the Sun. When we look at the hour-hand on our watch, in essence we're watching it rotate with the spinning Earth – but usually at twice its speed, because most watch hour-hands go round every twelve hours, not every twenty-four. All we therefore need to do to *set* our watch, is to measure the rotating Earth and transfer that measurement to the watch-hands. We could use the Sun as our reference point but it's very big, so we wouldn't get a very accurate result if we measured its position in the sky. Instead we usually use the stars. They appear to rise and set just like the Sun, but they're smaller, further away, and easier to observe.

So we get a telescope, fix it in position, choose some nice bright stars to watch out for every night and time them as they appear in (or 'transit across') our telescope sight. If our telescope can swing up and down (but not side to side), we can observe different stars through the night to get a more accurate time fix. For two successive transits of the same star we look for it at about the same time on the following night. What the astronomers at Greenwich had was high-precision, expensive, accurate transit telescopes mounted on very solid foundations. All were kept perfectly in order by careful measurements and adjustments and were monitored night after night by the astronomers, who both observed with them and used the observations to correct very good clocks next to the telescope. Night after night of careful, patient observation was the basis of precision astronomical timekeeping from the seventeenth century to the 1950s.

London watchmakers wanting to know the time had two options. One was to mimic the astronomers by using a miniature version of the time-finding transit telescope used at Greenwich. Some called it 'striking a transit', and it was difficult and time-consuming. Transit telescopes cost a lot of money, too, and without the patient attention and solid infrastructure available at a national observatory, the results could never be as accurate or reliable. But they could be good enough and it was done. Ruth Belville herself once acknowledged, 'One well-known London firm used to strike a transit every day for themselves, and share the result with several other firms.' John Harrison, ever the maverick, had used the chimney-stack of a nearby house as a transit reference, lining it up with the edge of one of his own house windows and waiting for the elusive stars to appear and disappear. But this was not a method for everyone, and the watchmakers had an alternative, which brings us back to Airy's new rule about receiving visitors.

You could get a good-quality pocket watch – probably one of the firm's own – and send one of the shop apprentices to Greenwich with it each week. He could knock at the gates, ask the astronomer for the time, note the difference between proper Greenwich time and the reading on the shop's watch, and then scurry back to work as fast as possible with the corrected watch. John Arnold certainly did this. Once, in 1782, he was involved in a slanging match with a rival maker, Thomas Wright. Arnold implied that Wright was not able to make a watch himself and did not even know what the time was. Wright retorted, 'Mr Arnold, it does not signify whether I can make a watch or not, I don't fear getting plenty of employ at mending thine, and if the watchmakers do not know what o'clock it is, they can know by going to Greenwich for it as thee does.'

Cattiness aside, Arnold was not the only one who sent to Greenwich for the time. One of the assistants working for Benjamin Lewis Vulliamy, another top maker, recorded journal entries such as, 'Went to the Post Office, afterwards to Greenwich, called with time to Mr Barraud, Grant & Dent.' This was 1835, just before the London & Greenwich Railway opened. Steamboats plying the river Thames offered the fastest route to Greenwich, but still the round trip took several hours. In every other respect, Vulliamy, Barraud, Grant and Dent might be deadly rivals, but all four were content to work together in a time-distribution cartel because knowing the time at Greenwich was the bedrock of their business.

But that was under John Pond's easy regime. George Airy wasn't so happy. His new rule limiting visits to Mondays only must have helped, but there was an alternative option to reduce still further the number of times the gate bell was rung. Instead of the watchmakers coming to Greenwich for the time, perhaps Greenwich time could go to the watchmakers. If the workshop assistant was spending hours every week journeying to Greenwich and back just to set the regulators right, he was not performing the myriad technical tasks required back at the firm's workshop for the bulk production of watches and chronometers. Their contact at the Observatory, the man who superintended not only the transit telescope and the time ball, but the chronometer department too, was John Henry Belville. Was there any chance, they must have asked informally, that he could bring a corrected chronometer to them every week? The answer was yes – the Astronomer Royal himself was as keen on the idea as the London watchmakers.

Belville had a close relationship with the watchmakers up in London but he was a very busy man. One option would have been to take a pocket watch into London himself each week. But in 1835, when Airy joined the Royal Observatory, he listed all the duties of his assistants. 'Mr Henry . . . manages the observations with the Transit, the reductions of the Transits, the rating of the Clocks, the comparisons of Chronometers belonging to the Royal Navy, or on trial for purchase by the Government, and the dropping of the Signal Ball at 1h mean time every day.' Quite a workload, and he had to be on-site every day and night to keep on top of it. He might have liked the idea of journeying around the city with his corrected watch, but there just wasn't a chance he could spare the time (as it were). But *sending* one down was a different matter. Pop it into a messenger's bag, give him a list of names and addresses and wait for it to be sent back at the end of the day or week. This was much easier, and it would keep his chief happy if there were fewer interruptions back at Greenwich so they could all get on with their work.

Belville was certainly not afraid of hard work, and it was not just in astronomy and timekeeping. From the first year of his stay at the Royal Observatory until just before his death – a period of forty-five years –

he meticulously noted detailed weather observations in a series of weighty journals. Every day he would record the readings from his own meteorological instruments in neat tables, alongside written notes on the state of the weather that day. It became obsessive. Three times a day he would make entries in his books, stopping only very occasionally for a day or two, when he was dealing with bereavements. In fact, he was more likely to pause for his pet than for any of his wives. Neither Sarah nor Apollonia ever got a mention, but he took the trouble to sketch his dog, Dash, in 1836, and on its death on 13 June 1842, Belville inscribed a solemn entry, framed in black. 'At a quarter past 7 AM poor little Dash died, his age probably 12 years. He died of asthma, perhaps the intense heat hastened his end. Alas, poor Dash!!' Belville himself suffered from asthma, but perhaps his wives did not.

The weather journals make for dry reading, apart from the Dash entries, a poem and a couple of shopping lists. But there is one tantalising five-line pencil note in the endpapers of the journal covering 1835 to 1840. It is almost illegible compared with the neat notes and careful numerals of his weather reports, but sometime between those two dates, John Henry Belville scribbled down a list of five names and addresses in London: 'Mr Eiffe, 1 South Crescent, Bedford Square. Mr Muston, 18 Red Lion Street, Clerkenwell. Mr Porthouse, 10 Northampton Square, Goswell Street. Mr Hewitt, 12 Upper Ashby Street, Northampton Square. Mr Hislop, 96 St John Street Road.' Five men, five addresses. Who were they? They were leading London chronometer-makers. Why had Belville scrawled them into his weather journal? Was he planning to send them something? It might simply be that he was sending back some chronometers they had lodged with him for testing. But perhaps this was a list of customers subscribing to his new time-distribution scheme.

He had plenty of eager clients and a chief who was happy to support the project. Now, he needed a suitable chronometer that could be sent into London every week. This was when a steely mechanical heart started beating in our story. Belville chose a pocket chronometer (no. 485) made in 1794 by John Arnold. Legend has it that the gold-cased watch was made for the sixth son of George III, Prince Augustus Frederick, but that he thought it was too clumsy (he is supposed to have said it looked like a warming pan) and returned it, whereon John Belville bought it for his time-distribution service. There may be an element of truth in this, because the prince (who was also Duke of Sussex) was President of the Royal Society and a member of the Royal Observatory's Board of Visitors. He would therefore have toured the Greenwich facilities every year and met the assistants, so perhaps he did a deal with John Belville over the watch. The gold case would have to go, though. In 1840 the watch was rebuilt with the latest chronometrical gadgetry and put into a silver case. Newspaper reports nearly seventy years later offered Ruth Belville's explanation. To one paper, she said, 'Mr Belleville substituted a silver case for the original gold one because his curious profession took him occasionally into the less desirable quarters of the town.' To another, her words were, 'When my father bought it he had the heavy gold case replaced by a silver one, as he had to visit many London slums, and thought it would be safer if it looked less imposing.' Less imposing, maybe, but it was in constant use by his family for well over a century.

Now that Belville had set up a chronometer time service, things at the Observatory must have been more to Airy's satisfaction, but the 1840s proved a mixed bag in Belville's own life. In 1840, he and Apollonia moved again with their daughter, Cecilia, to a little brick cottage in nearby Prior Street, Greenwich, which housed a varied population, including the likes of tailors, shipwrights and book-keepers. One day in December 1843, John was at work when he was approached by James Glaisher, one of his staff working in the Observatory's magnetic and meteorological department. William Thackeray, an assistant at the Observatory at the turn of the twentieth century, recounts the story. 'Glaisher, a young assistant, asked Belville (his chief at the time) for a day's leave. On the latter enquiring, "for what purpose," G's reply was "To marry your daughter!" Whereupon B, not unnaturally, "flared up."' It might just be one of those stories that get passed down over the years but Belville would have had ample reason to flare up, over and above indignation at the abrupt way Glaisher asked for his daughter's hand in marriage. Glaisher was not so young. He was 34 years old; Cecilia Belville, John's daughter, was just fifteen. The newly-weds moved to Blackheath and John's relationship with his daughter broke down; his working relationship with Glaisher following the marriage can barely be imagined.

The changes kept coming. In 1844, John and Apollonia moved to 9 Hyde Vale Cottages, a large town-house surrounded by merchants, ministers and company secretaries. They must have hoped things were settling down as they were moving up. But all the time Belville was dealing with house moves and family breakdowns, the Astronomer Royal back at the Observatory was planning a huge new project that would change his entire working life. In 1849 Airy revealed his latest idea in time distribution: time by telegraph.

The plan was simple in concept. Airy envisaged one single clock keeping time for the whole Observatory. Like all observatory regulators, this one would use a swinging pendulum. But unlike existing clocks, which connected the pendulum through a train of gear-wheels to the hands on the dial, his new idea was for the pendulum to operate an electric switch every time it swung. The electric switch operated a battery circuit – on and off, once every second, and each time the switch closed, a pulse of electricity from a battery would flow through the circuit. The clever part was that a whole string of simple electric clock dials would be attached to the circuit – as many as were required – and on receiving an impulse down the wires, their hands would advance by one second. The result on each dial looked little different to the hands on a normal wind-up clock: the second hand clicked forward every second, the minutes and hours following at their own rate. But in each of these electric clock dials there was no spring to wind up, no ticking pendulum, no big train of wheels to maintain. The dials took their power from the electric wires, they took their timekeeping from the big pendulum clock, and they were simple. And here was the crucial development: every dial on the circuit would show the *same time* as the single pendulum clock. There could be no variance between clocks in different rooms; no confusion over which clock was right.

Airy's plan was bold. Anywhere an electrical wire could go, the Royal Observatory's time could go too. It need not remain inside the Observatory's walls. Time could be exported instantaneously from Greenwich. It just needed a network of electrical wires, and conveniently, at just the same time, telegraph wires were being strung along every railway line, across every urban street. You can see where Airy was going. As well as telegraph messages, these wires could carry Greenwich time. Not one pulse each second, as in the Observatory's own internal network – this would simply clog up the system. No, these telegraph pulses would be hourly or daily time *signals*, helping people far from the Observatory set their own existing clocks just as John Belville's corrected chronometer did.

Airy explained his scheme. 'I fully expect in no long time to make the going of all the clocks in the Observatory depend on one original regulator. The same means will probably be employed to increase the general utility of the Observatory, by the extensive dissemination throughout the kingdom of accurate time-signals, moved by an original clock at the Royal Observatory; and I have already entered into correspondence with the authorities of the South Eastern Railway (whose line of galvanic communication will shortly pass within nine furlongs of the Observatory) in reference to this subject.'

'Galvanic communication' was one of the terms used for the electric telegraph at the time and, in this brief reference to 'general utility', Airy's strategy is clear. In a world of changing priorities and new technologies, institutions like his needed to adapt to survive. The Royal Observatory had been founded in 1675. Its original function was to map the stars as part of the great longitude project for safe navigation at sea. But that had been completed. Airy needed a new *raison d'être*. With the electric telegraph and his 'original clock', Airy could provide the time for Britain.

In 1851, he inspected a system of 'sympathetic clocks' made by Charles Shepherd, on display at the Great Exhibition, held that year in Paxton's vast Crystal Palace in Hyde Park. It was an impressive sight, as the *Illustrated Exhibitor* explained: 'Mr Shepherd's electric clock claims especial notice. The hands are outside the building, but the mechanism is in the south gallery, fifty feet below . . . The whole of the works of this great clock are kept in motion by a series of powerful electro-magnets; and by means of an immense coil of copper wire, other clocks in the Exhibition are kept going.' Airy liked what he saw so much that he commissioned Shepherd to build a similar system for the Royal Observatory. In 1852, after engineers had laid telegraph lines across Greenwich Park to nearby Lewisham railway station, the new Shepherd 'motor clock' was set ticking. Small dials in rooms around the Observatory were connected up to the new master timekeeper, and a large white 'sympathetic dial' was mounted outside the Observatory's gates, displaying Greenwich time to the public – the first clock to do so as it still does today. The Greenwich time ball was wired up to the new electrical system, freeing up an assistant or two every afternoon, and the one-second time pulses began speeding down the wires to Lewisham where they joined the trunk cable to London Bridge station. There, they ended at a final 'sympathetic clock' next to a switching unit which sent hourly, twice-daily or daily time signals around the network: to the railway companies, the Post Office, and even to the Electric Telegraph Company's headquarters on the Strand, opposite Charing Cross station, where they operated a time ball on the roof. Airy was pleased. This was a modern, scientific project that was worthy of time and attention from his great institution – and was a great way to ensure its ongoing survival. After all, who could ever switch off the signals once they started beating time for the nation? As Airy observed, 'I cannot but feel a satisfaction in thinking that the Royal Observatory is thus quietly contributing to the punctuality of business through a large portion of this busy country.'

But if the Observatory was helping a 'busy country', it hadn't forgotten its existing London clientele. The Shepherd clock made the headlines but John Belville kept on sending his Arnold chronometer up to the London watchmakers each week. New technologies don't simply replace old ones. Usually, they just add another layer of complexity to our lives. The electric pulses shot along the wires from Shepherd's galvanic clock to London Bridge and beyond but, like all new technology, it took a long time to bed in. All the while, John Belville continued to send time the old way – the slow way – the reliable way.

It's odd, though. Most of this story has come third-hand from John's daughter, Ruth. Neither Belville nor Airy referred to the time-supply business. Ruth always said that John carried the Arnold chronometer *himself* around his customers but, as we saw earlier, that would have been very difficult with his workload. It's just possible that Ruth was mistaken and things weren't entirely as she understood. In 1938, one of the Observatory assistants, Henry Hollis, was delving into the institution's history. He wrote, 'I have been told . . . that Belville started the business of carrying the time to London; but Atkins of the firm Brockbank and Atkins said once that this is wrong. The chronometer-makers appear to have another story. Atkins is dead now – I regret I did not ask him.' Hollis's interest may have been sparked in 1926, when he received a letter from a French chronometer-maker recounting a story that had gone around the French horological trade in the 1850s involving an old London clockmaker selling the time. Hollis was intrigued. 'It is <u>tradition</u> in the Observatory that Belville established the business, and the ladies who carried it on were certain-ly of his name, but if there is some other story I would like to find out.' Was it somebody else – or, more likely, did the arrangements change over the twenty years after Airy arrived?

The French story had come from an 1857 article in a horological journal, *Revue Chronométrique*. A London clockmaker, explained the author, had become too old to carry out his ordinary work but in return for a small payment now provided for his fellow workers the service of bringing them the time from the Observatory to regulate their clocks. There is no mention of a name, but did this old French report reflect the Belville story? Perhaps. It might even help us under-stand why Ruth always thought John carried the watch himself.

In 1851, Belville's second wife, Apollonia, died at their Hyde Vale home. Later that year, on 22 December (at age 56) Belville married his third wife, a 39-year-old Suffolk-born teacher, Maria Last, in the parish church of St Andrew, Holborn. Maria lived in Cross Street, off Hatton Garden. Cross Street also housed St Andrew's Parochial School, built in the seventeenth century by Christopher Wren (as the Royal Observatory itself was). Perhaps Maria was teaching there; perhaps John met her while visiting customers in Hatton Garden (there were several watch- and instrument-makers there at the time). On 5 March 1854, back at home in Hyde Vale, their daughter Ruth was born. John Belville had done pretty well, all things considered: three wives, the last one seventeen years his junior; children by each marriage, the youngest born when he was 58. But by this time his own health was in steep decline. His asthma was bad and he had developed an abdominal tumour. He had long feared the effects of the air at Greenwich on his condition, once asking John Pond, 'whether the air at this particular place . . . was good, and conducive to health'. Pond, apparently, had said yes, 'saying that all the Astronomers Royal died at an <u>advanced age</u> and also that no assistants died while in service at the Observatory, that although they complained of the disagreeable things <u>about the Observatory</u> they still kept to it.'

But the air in Greenwich, below the Observatory, has always been some of the worst in London for asthmatics and can't have been good for Belville; by 1854, he was finding it more and more difficult to 'keep to it'. In August, aged 59, he asked George Airy whether he could retire because of his health. Airy refused. 'Of Mr Henry's own comfort he is himself the best and proper judge, but I can hardly imagine that his happiness would be increased by quitting office now or soon. Indeed I think he would find himself very incomfortable.' Nevertheless, Airy had no wish to see Belville suffer unduly. He wrote to his chief assistant, Robert Main, telling him to devise a new work schedule with Belville. 'Mr Henry may possibly desire, frequently or generally, such arrangements as may relieve him somewhat in regard to over-strict attendance and bodily labour. I have told him that we have every disposition to meet his wishes on these points (an attention to which his long services and seniority fully entitle him).'

Belville was relieved and grateful, writing a frank and emotional note to the Astronomer Royal expressing his 'heartfelt thanks' for the favour Airy had paid him. Maybe this makes sense of the French report in 1857 of the 'clockmaker' who had carried the time around London as his age had rendered him unfit to carry out his ordinary business. John Belville may have started the time business by *sending* the time into London. Then, by the 1850s, when he was finding it difficult to work full-time at Greenwich because of his illness, perhaps he did indeed begin carrying the chronometer himself, rather than sending it by messenger. The change of air might have done him some good, and like anybody he had his pride; he needed to feel he was still useful. And it would have helped Airy square things with the Admiralty, who were not renowned for their generosity towards civil servants. Ruth was only born in 1854 so she did not have first-hand experience of the earlier years of her father's work, which could explain why she might have got the detail wrong.

But if John was the old man in the French journal article, that still does not explain Brockbank and Atkins's 'other story'. Perhaps we will never know exactly how John ran his business, but, as Henry Hollis said, it was the start of a tradition. Belville was not going to see it blossom, however. On 6 December 1855, after twenty-two years of operation, the timber mast supporting the Greenwich time ball on the roof of Flamsteed House broke in a winter gale. It was just before one o'clock, and the ball was being raised for the daily signal. The five-foot diameter wood-and-leather ball, together with its mast, came crashing down into the courtyard. On-site to witness the aftermath was Hubert Airy, George Airy's 17-year-old son, who quickly sketched the scene in a watercolour that still exists in the Observatory's collections. The episode was a portent of what was happening in the time ball supervisor's own family life. The world of Maria and Ruth Belville was also about to come crashing down.

Chapter 2

MARIA BELVILLE'S TIME SERVICE

{ 1856–1892 }

The Royal Observatory gates, about 1870.

9 HYDE VALE COTTAGES, GREENWICH, SUNDAY 13 JULY 1856. John Belville's health had taken a rapid turn for the worse. He now lay in bed, at home, exhausted and in great pain. His long and severe illness was reaching its conclusion, but in his final few hours a reconciliation took place. A family matter that had preyed on his mind for a long time reached a kind of resolution. His 28-year-old daughter, Cecilia, with her husband, James Glaisher, now 47, were by his bedside as he lay dying. He had hardly spoken to Cecilia in the thirteen years since their shock marriage but now she was here, and so was James, and he was glad to see them. Comforting words were exchanged and at five o'clock in the afternoon John Belville died. He was just eight days short of his sixty-first birthday. He was buried five days later in an unmarked grave in Lee churchyard, near the tomb containing John Pond.

The Astronomer Royal, George Airy, was abroad on an expedition with his 19-year-old son, Wilfrid. First they had gone to Paris before travelling through Italy to Sicily. Airy's journal records his whereabouts on the Friday before Belville's death: he saw the meteorological observatory on Italy's Mount Vesuvius, a sight that would have thrilled his weather-obsessed assistant. While he was away, his first assistant, the Reverend Robert Main, had been left in charge of day-to-day affairs at Greenwich. On administrative matters he had no reason to write to the Astronomer Royal but on Monday 14 July, as Airy journeyed through Italy, Main penned a short letter to his chief, informing him of Belville's death. 'He has been a great sufferer for a long time, and the change cannot but be gain to him. It is gratifying to know that his daughter Mrs Glaisher who was so long estranged from him, and her husband, were both by his bedside when he departed, and that he expressed his satisfaction at seeing them.' Main's letter was waiting for Airy at Marseilles, on France's south coast, upon his arrival from Italy on 29 July.

Airy and his son were both ill themselves, confined for three days in the French port 'from a touch of malaria'. Airy was saddened by what he read. He wrote back to his first assistant. 'The letters which I found here informed me of the loss of the oldest and one of the most faithful servants of the Observatory – one who connected the ancient and the modern states of things.' And yet Airy was unable to switch off his bureaucratic brain. In the very next sentence, he advised Main of the administrative procedures to follow in the event of a death in service. 'You have probably remarked the official steps usually taken (see "Appointment & Reward of Assistants", A.4 I think, or the bound volumes on the highest shelf of the right-hand compartment of AR. MSS) to announce this to the Admiralty,' and, following that, he instructed Main to ask Belville's widow, Maria, for various documents about the Observatory that would now be in her possession. She should be invited to give them to the Observatory, as they would be 'totally useless in every other place', and he concluded his letter with an account of recruitment procedures now that a vacancy had arisen. He returned to Greenwich on 5 August, ready to make the changes.

It might be easy to assume that Airy cared little for his assistant and was more interested, perhaps, in the rules and regulations that he could now press into service following Belville's death. Yet this would be wholly unfair. Airy was a practical man for whom rules and regulations mattered. He was trying to create the order and discipline he felt had been lacking under his predecessor. But Airy certainly had the capacity both to feel emotion and to act emotionally, and the death of John Belville hit him hard. Speaking at the annual visitation day on 6 June 1857, Airy informed the visitors of the changes in staffing over the previous twelve months. 'Within the last year we have lost by death the Senior Assistant, Mr John Henry, the last who remained from Mr Pond's establishment, and one of the most faithful and zealous of my coadjutors.' High praise indeed from Airy in this formal setting. What he found more difficult to deal with was the situation that Belville's wife, Maria, and their two-year-old child, Ruth, found themselves in following the loss of their breadwinner. Maria needed an income and Ruth needed security. A family friend offered money to pay for Ruth's education but Maria had to refuse it: as Ruth recounted later, 'she was in possession of her own small income and did not wish me to leave her.' This small income came from Maria's work as a teacher. But if Maria was in a difficult position, where did that leave George Airy?

The Observatory owed John Belville a favour, thought Maria, given the lifetime of work he had given in its service. Three weeks after his death, she wrote Airy a long and emotional letter. She was in a fix and she was pleading for help. She told Airy that her 'dear departed husband . . . has more than once said "When I am gone I am sure that Professor Airy will assist you as he has done me that is to say plead for you as for one of his own family to obtain a pension from the Admiralty which my long services justly entitle me to."' Their daughter Ruth, John's last child, was only two years and five months old, Maria told Airy. Her future needed to be provided for. Maria wanted to invest the 'few hundreds' that John had left behind for Ruth's benefit, but she could only do this if she, Maria, had an income of her own. For this, she needed a Government pension but she was clearly aware that this was a big request. She wanted to offer something in return, some collateral that Airy could use to argue her case with the Admiralty. On offer to Airy, if he was interested, were John's meteorological diaries, a weather journal by the previous Astronomer Royal, a third manuscript from the seventeenth century plus a mixed bag of meteorological papers and measurements. There were several scientific measuring instruments, including a thermometer by prominent instrument-maker, John Bird. 'The child being a girl cannot utilise these,' said Maria, otherwise she would not have considered parting from them. Finally, Maria asked Airy what chance she had of getting the Admiralty to engrave John's name on the tomb containing John Pond. Not a chance, it seems. Ruth Belville recounted a story many years later to Observatory staff: 'in the disused closed Church Yard of St Margaret's Lee there is an old vault tomb that belonged to the Admiralty – Mr John Pond was interred in this vault – and my mother sought permission from the Admiralty that my father might receive burial in this also. But this permission was not granted – so he lies in a grass-covered grave along the side of the vault – so after all he is not so far from the Observatory he loved so well.' Ruth may have been mistaken. Her father's body was interred five days after his death – hardly time for Maria to correspond with the Admiralty on permission for a burial. Perhaps she had already asked while he was still alive, or maybe all she ever wanted was for his name simply to be marked on the tombstone. It never happened.

But the sale of the books and instruments seemed a more viable proposition. Airy mulled the matter over and replied a few days later with his initial assessment of the situation. 'There is not the most trifling possibility that a pension would be granted to you.' In the past, he said, he had tried to obtain pensions in related circumstances but his applications had been 'flatly refused'. On one occasion, he had managed to secure a small one-off payment of £20 following the death-in-service of an assistant, but the next time he tried it, 'they refused to give anything.' Things got worse. Every other member of staff on the Observatory's payroll made a monthly payment from salary towards a 'superannuation fund' to cover pension costs in later life. Yet, perhaps because he had been on the books long before Airy arrived with his new administrative regulations, 'Mr Henry's salary alone was not charged with this deduction.' Things were not looking good, either for a regular pension or even a small ex-gratia payment. Airy concluded, 'I do not think, therefore, that there is the smallest chance that any money could be obtained.'

A few days later, though, Airy had been doing some more thinking. The meteorological papers might form part of the answer, although he didn't hold out much hope. If he could transact a simple, one-off sale, perhaps they could secure enough money at least to help the Belvilles on their way. He wrote to Maria asking her to send the weather journals and manuscripts to him for inspection, concluding, 'if you have any precise idea as to the recompense which you would expect for them, you would perhaps have the kindness to state it.' He could then, he said, forward the price to the Admiralty and ask them to make a payment in these unusual circumstances, although he could not make any promises. Maria was grateful to learn of this turn of events, and sent the journals to the Observatory, but she would not be drawn on their value. 'I find it impossible to affix a price . . . I should much fear to be too exacting . . . but his often expressed wishes that they might be retained at the Observatory and his long and intimate acquaintance together with his entire confidence in you fully justify me in praying you sir to name their value yourself.'

Airy took a few days to look through the documents and consider the situation. It was worth a try, he concluded, and wrote to Maria. 'You are aware, from what I have already said, of the extreme difficulty of successfully urging a claim on the Admiralty: and that in any case the most that I can do is to submit a matter for their decision.' He was trying hard to keep her expectations in check. 'Now I think I may submit this to them – that in consideration of the long and faithful service of Mr Henry, and in recognition of the present of the Meteorological Manuscripts a grant of £100 should be made to you.' Maria Belville was delighted. The Astronomer Royal wrote to the Secretary of the Admiralty. The letter lavished praise on 'Mr Henry': 'there never was a public servant more faithful, or more devoted to his office . . . Mr Henry was also a man of gentlemanly spirit and gentlemanly manners: and these things are important in an institution where, the salaries being generally low, we have some difficulty in finding Assistants above a very low rank.' In light of this, coupled with the fact that his unique and extensive weather journals were now at the Observatory, Airy laid out his claim. 'I now submit to their Lordships whether, in accepting these manuscripts, and in recognition of the long and faithful services of Mr John Henry, their Lordships may be pleased to grant £100 to Mrs. M. E. Henry Belville.'

The reply came back swiftly: 'I am commanded by their Lordships to inform you that as it is not the practice to provide for the widows of Civil Servants, they therefore are unable to render assistance to Mrs Belville.' Airy wrote to Maria. 'I inclose a copy of the reply of the Secretary of the Admiralty to my application in your favour. I am sorry to say that it puts an end to all hopes. I was so well informed of the unwillingness of the Admiralty to make any such grants that I should not have thought of applying except under the strong claims of Mr Henry.' And that was that. Maria wanted the weather journals back, naturally. She could get money for them elsewhere, although she very much regretted that they would not stay at Greenwich. She told Airy, 'I should have been much gratified by their remaining in the Observatory at any price, such having been his expressed wish – but I cannot forget his earnest injunction to look after the interests of "poor little Ruth".' Airy was utterly dejected at the defeat. 'The Meteorological manuscripts shall be sent down to you. I regret exceedingly the decision of the Admiralty which leads to this.'

Application to have her husband's name marked on the Admiralty tomb: rejected. Application for a pension: rejected. And application for a one-off payment for the weather journals – her husband's life's work – rejected. But was this really an end to all hopes? Not quite. Maria Belville had a further proposition for the Astronomer Royal.

The idea was put to Airy six weeks after John Belville's death. 'I am encouraged by your goodness to advance another petition. Being engaged to take the Greenwich time to 67 of the principal chronometer-makers in London I have to request admission once a week to the clocks in the Observatory in order to test my own regulator – it would inspire those who have taken up the widow of their esteemed friend with additional confidence if you could accord me this favour.' Maria wanted to take over the time-supply business. According to Ruth Belville, writing many years later, Maria was already well known at the Observatory and had been involved in the chronometer work for four years. Ruth recalled, 'Owing to the illness of my father in 1852 my mother began taking the chronometer and checking the errors in his stead. First to the most unimportant firms, later on to the others, and at the death of my father in 1856 she was requested by nearly all the firms to continue the work.' It would certainly have pleased Airy that she had worked out a way to make some money, and while there is no record of his reply, he was happy to continue letting her in. Maybe it was even his idea that she formalize the whole matter in writing. She was never officially connected to the Royal Observatory, but this was the formal establishment of Maria Belville – the first 'Greenwich time lady'. And yet, despite Airy's support, it almost ended as soon as it had begun. He was happy to let her check the time each week – but only by his strict rules. Maria Belville was about to learn a lesson she would never forget.

It was Saturday morning, the first day of November 1856. Maria was coming to terms with her rough treatment at the hands of the Admiralty, and now she needed to concentrate on earning a living for herself and her young daughter. So here she was at the Royal Observatory gates with her husband's trusty chronometer, ready to collect Greenwich time and take it up to her customers in London. The Observatory was busy that day. For the past two months, while Maria had been negotiating with the Astronomer Royal over Admiralty pensions and weather records, workmen had been laying foundations for a new telescope building to the south-east of the existing complex, and George Airy had been keeping a close eye on their work. As she arrived at the Observatory that Saturday, Maria discovered that the main gate was ajar. She assumed it had been left open by the workmen. Since she could see that the porter was in his lodge, she passed through the open gate into the front courtyard. The business of the day was transacted quickly; her chronometer was checked against the Observatory clocks and she was ready to make the journey into town. All that remained was to bid good-day to the porter, who by now was sweeping out the long passageway linking the telescope rooms to the Astronomer Royal's residence in Flamsteed House. As she visited the Observatory that winter morning, the last thing Maria Belville wanted to do was to incur the wrath of the Astronomer Royal. But that is precisely what she did.

The following Monday, a letter arrived at the Belville house in Hyde Vale. By now, Maria was familiar with the astronomer's distinctive handwriting; she would have known at a glance that the envelope contained a letter from the Observatory chief. She tore it open and began to read. 'The gate porter of the R. Observatory . . . reports to me that you have entered the Observatory inclosure by unlocking the small entrance gate, without ringing the bell. I am sure you will see after a very short consideration that this cannot be permitted.'

Airy was furious. His instructions had been clear: anyone who wanted access to the Observatory had to be let in by the porter, and this applied 'only to persons whom I recognize as regularly attached to the Observatory'. What was this woman doing by sidestepping his orders? 'If the key which you have used was originally derived from the Observatory, it ought to be returned at once. If it is a key not originally intended for our lock, it is clearly improper to use it in our lock.' He finished his handwritten letter with a terse instruction:

'Have the kindness in future in all cases when you desire admittance to ring the Porter's bell.'

Maria was shocked and alarmed. She had only been in the business for a few weeks and she had angered George Airy, the arch-administrator who only allowed her to visit each week out of loyalty to her late husband's memory. He had accused her, in essence, of breaking and entering; how could this have happened? She dashed off a reply the same day and hastened with her letter to the Observatory. She needed to set the record straight. 'I am extremely sorry at having unwittingly transgressed any law of decorum on Saturday,' she wrote. She explained the circumstances, how she had found the gate open, how she had assumed it was the builders who had done so. 'I enclose the receipt given me by Mr Glaisher on Tuesday the 22nd July when I surrendered every key I possessed belonging to the Observatory.' She was indignant at Airy's false accusations. 'I could not have entered without finding the door open as it could never have entered my mind under any circumstances to apply a key to a lock of the Observatory', she wrote. 'I did indeed wish to retain the key of the George Street Gate but one of the Observatory never entered my thoughts.'

George Street (now King George Street) led from her Hyde Vale residence to a pedestrian gate into Greenwich Park, whose hours of opening by the park authorities were restricted. Observatory assistants often forgot their own park keys and had to climb over the gate – not an option open to a middle-aged woman with a small child in tow. This seemed a reasonable request. But a key to unlock the Observatory's own perimeter gate? Never. Airy was quick to back down. He wrote back, 'The statement in your letter received this morning is quite sufficient. The keys to which Mr Glaisher's receipt applies were duly received by me: perhaps his receipt had better remain in your hands.' Maria Belville could breathe a sigh of relief. In fact, she was so relieved at the outcome of this difficult exchange that she felt the time was right to involve the Astronomer Royal in the third phase of her plan to secure income for her daughter. She was about to call in some favours and Airy was the man to help her.

Maria Belville was, by profession, a teacher. Her main subjects were modern languages, although she knew enough mathematics to help her husband in his meteorological observations for several years as his health declined. Her intimate contact with the work of the Royal Observatory put her in a good position. A sound working knowledge of French and German was vital in the astronomical sciences, where so much work was international and so many astronomical expeditions multilingual. George Airy expected his brighter assistants to learn French and German in their own time, and such extramural activity saw them well placed when vacancies for promotion became available. Maria Belville was there to assist – and by helping prominent figures learn foreign languages she also made useful contacts in her own right, contacts that would prove useful in her wider quest to secure a future for Ruth.

One such pupil was William Marriott, who joined the staff of the Observatory in 1869 as a 'supernumerary computer' in James Glaisher's magnetic and meteorological department. He stayed for three years, apparently a quiet and methodical man who while at Greenwich was elected a fellow of what became the Royal Meteorological Society. He also had his eyes on the Society's own job opportunities, and in 1872 he successfully applied for their post of assistant secretary. He stayed there for forty-three years. In his application, Marriott noted that he had a working knowledge of French, enabling him to translate written work for others. This had come from Maria Belville, and it was Marriott's work for the Meteorological Society that finally led to a suitable permanent home for Belville's weather journals. Many years after the Admiralty had refused to pay for them to be housed at Greenwich, Maria passed the manuscripts to Marriott and in 1882 the Society bought them, keeping them to this day in the National Meteorological Archive near the Met Office in Exeter. Ruth recalled the episode some seventy years later: 'I remember as a child some very ponderous leather bound books which contained observations and notes on the weather written by my father. A Mr William Marriott who entered the R.O. as assistant I believe in the eighteen-sixties, had received from my mother a thorough knowledge of French in which language she was proficient. Mr Marriott spoke of my father's books to the Meteorological Society. They came and inspected them and my mother accepted an offer for them – and I suppose they are still in the possession of the Meteorological Society.'

But while teaching had brought Maria a small income independent of her time-supply work, and went on to secure her a one-off payment for the weather books, she felt there was more discussion to be had with the Admiralty in the matter of a pension. Her teaching had provided her with another contact – and this one was a heavyweight. Late in November 1856, after the incident over the Observatory gate key had died down, she explained the situation to George Airy. 'I have been so fortunate as to obtain through the medium of the Count Bouët de Willaumez, the inspector of the French fleet in Greece and to whose wife I had the honour of giving lessons for many years, a letter of introduction to Admiral Dundas.' The Crimean War, in which Britain fought alongside France, Turkey and Sardinia against Russia, had ended with the Treaty of Paris eight months before

Belville wrote her letter. Louis-Edouard Bouët-Willaumez was a French rear-admiral who took a prominent role in the war. James Whitley Deans Dundas had been a Liberal member of parliament for Greenwich in the 1830s and again between 1841 and 1852, in which year he was appointed commander-in-chief in the Mediterranean and elevated to rear-admiral. Dundas too played a prominent role in the Crimean War. These were good contacts for Belville to cultivate.

She picked up the story with Airy, explaining that Dundas had 'kindly promised to present himself and support with all his influence with the Lords of the Admiralty my petition for a pension in consideration of Henry's long service.' Belville had drafted the wording of a petition and now asked Airy if he would look it over and sign it. She knew that she was pushing her luck. 'Were I alone in the world I would not be this importunate but when I look at his child and remember how inexpressibly dear she was to his loving heart I feel myself emboldened. It would be so hard to leave her unprovided for.' The trouble was, she had never written a formal petition before, and she had no idea how best to frame one. Airy knew Dundas, having worked with him on a naval engineering project in the early 1850s. Could he help?

Airy read Maria's draft and hated it. Too long. Too much repetition. No logical order or sequence. It needed a complete rewrite but he would never be so blunt as to say that. She needed coaching, that was all. The roles were reversed: Maria was now the child, Airy the patient teacher. 'Madam, the first draft of a memorial is of course liable to some corrections, and I think you will find that that which you have sent . . . may be put in a somewhat more orderly shape.' Airy, Britain's most senior astronomer and the man in charge of one of the country's foremost scientific institutions, was taking the time to offer writing lessons to Maria Belville. 'I can give no suggestion except that you should read it over carefully several times, and while on the one hand you take care to express all that you intend, on the other hand you should remove repetitions and arrange the order of the representations.' Of course, he shouldn't really be involved at all: 'I am very busy, and am myself unable to enter into the details. I do not think it would be proper for me to sign such a paper, because I have communicated officially with the Admiralty. But I think that you would be quite justified in referring to my communication.' Airy ended by proposing a form of wording that might indicate his tacit support for her claims without saying it in so many words, and he signed off, lesson over, homework set. Was Maria successful? There the Observatory correspondence ended, and suffice to say that Maria continued selling the time and teaching her pupils for the rest of her working life, so it is a reasonable assumption that the Admiralty didn't budge.

By 1858, Maria and Ruth Belville had given up 9 Hyde Vale Cottages and moved around the corner to an imposing Georgian town-house in Crooms Hill, an old street that runs up the western boundary of Greenwich Park and even closer to the Observatory for Maria Belville's weekly visits. Their neighbours were well-to-do professionals: doctors, solicitors, architects. To cover the rent on the Crooms Hill residence, which was arranged over four floors, Maria set up a private girls' boarding school which she ran until the 1880s. As she grew out of childhood, Ruth also worked there as governess from her mid-teens until the pair moved house again in the 1890s. She helped out from an early age in the time-supply work, too. A small child was no obstacle to the progress of the redoubtable Maria Belville, and Ruth often accompanied her mother on her weekly rounds. Like mother, like daughter. Time kept passing and time was passed on. Victorian

London grew and blossomed; industrial and technological developments transformed daily life and politicians reformed British society in myriad ways that depended on time for their success. Alcohol sale was regulated, with the introduction in the 1870s of two liquor licensing acts that imposed strict opening hours in public houses across the country. Working-time directives limited hours of work in factories. The railways spread to every town and even village, it seemed, and amid all this change, clocks ticked onwards.

In 1880, a law was passed establishing Greenwich time as the legal time for the whole of Britain. Before this, local time was generally the rule, and that could vary from the time at Greenwich by up to half an hour, depending on how far east or west the town was. The 1880 act regularized everything to the Greenwich meridian, the origin of the time that was the business of Maria and Ruth Belville. From now on, the act decreed to the nation at large that Greenwich *was* time.

So there was always work for the Greenwich time ladies. Their customer numbers stayed buoyant, their clients loyal, and one of them even tried a little product placement – although it does not seem to have had much effect. On 20 July 1882, after Maria had concluded a quarter-century supplying the time to the London horological trade, the watchmaking firm of Parkinson & Frodsham presented her with a silver-cased pocket chronometer, number 5076. There was no charge and it was never returned: a gift to the woman who brought them the time each week. But there is no record of the watch ever being used in the service of the time-supply business and no mention in any correspondence. Perhaps the women wanted to stay loyal to Arnold for a little while longer but the gift of a watch must have reminded Maria Belville of the inexorable passage of time. She was not getting any younger and London had its dark side.

Starting in 1888, the capital was rocked by a series of brutal murders of women around the Whitechapel area of the East End. Dubbed the 'Jack the Ripper' murders, these serial killings were sensationalized by the popular press, and nobody living or working in the city could have remained unaware of the tragic events that unfolded over the next few months. The murderer was never found, and the apparent impotence of the police in the face of killing after killing lent further weight to the public outcry. This was a frightening time for Londoners, and Maria Belville's weekly walking round took her right through the heart of the murder district. See the world through her eyes in 1888: you are a 76-year-old woman, you have become partially blind and you are carrying a valuable silver pocket watch through the slums of east London every week. You hear all the gruesome details about the first Ripper murders. The first killings took place just a few streets to the north of your main walking routes, which must have been shocking enough. But then events get even closer to home.

On Monday 1 October, the morning newspapers are full of reports of a further two murdered women, both found in the early hours of the previous morning. Perhaps your daughter will have read the newspaper reports to you over breakfast. One of the victims, identified as Elizabeth Stride, has been discovered near Berner Street (now Henriques Street), close to the London Docks, which is the first stop on your weekly time-distribution round. The second body, of Catherine Eddowes, has been found in Mitre Square, just off the Minories – which is your next regular port of call after the docks. Eddowes's corpse has been terribly mutilated. One of your own cousins, Thomas Lewes Sayer, works for the Town Clerk's office in the City of London Corporation and goes to view Eddowes's body at the city mortuary. Later, he recalls, 'It was a nightmare experience which I should not care to repeat. The terrible mutilation which had taken place indicated clearly that the work was that of a madman.'

She kept going, perhaps fearful, or perhaps with stiffened resolve. Nevertheless, in about 1889 or 1890, Maria gave up her girls' school in Crooms Hill and moved with Ruth to a brand-new property in Charlton, a residential district just east of Greenwich being heavily built up for housing along the fast North Kent railway into London. The two women now lived at Elliscombe Road in a much smaller building than their Greenwich town-house, surrounded by reasonably comfortable neighbours on ordinary wages. It made for a good home and they could concentrate on looking after each other – with the income of a single lodger, a civil servant, going towards the salary of their one domestic servant. Ruth, by this point, had taken up her mother's teaching work in earnest, her 1891 census return describing her as a 'Professor of French and Music'. But she would soon have to juggle her time between the teaching and a supplementary occupation to make ends meet. For Maria, a combination of advancing years, partial blindness and, doubtless, a weariness at the weekly struggle through a hostile London meant that the first Greenwich time lady was ready to call it a day. George Airy, knighted in 1872, had retired in 1881. On 2 January 1892 he died aged 90 at the White House, his home within sight of the Observatory at the top of Crooms Hill, ending a remarkable era at Greenwich. Five months later, Maria Belville, aged 81, retired.

Chapter 3

RUTH BELVILLE AND STANDARD TIME

{ 1892-1908 }

Maria Belville in the Daily Graphic, *1892.*

9 ELLISCOMBE ROAD, CHARLTON, 10 JUNE 1892. Ruth Belville, now aged 38, was at home with her mother in their little end-of-terrace house in Charlton. She was writing a letter to the new Astronomer Royal, William Christie, to ask if she could formally take over her mother's arrangement to visit the Observatory each week. She wrote in the third person. 'The late astronomer royal Sir George Airy for more than forty years had allowed her mother to take the time from the master clock in the Observatory but in consequence of her advanced age and partial blindness she is almost unable to come so far.' More than forty years. Much time had passed since Maria Belville wrote her own letter to the Royal Observatory asking permission to visit the clocks every week. Ruth continued: 'As some of the leading City and West End firms have kindly promised to continue taking the time during Mrs Belville's lifetime as they have every confidence in her chronometer, would the Astronomer Royal allow Miss Belville the privilege of taking the time from the master clock every Tuesday morning.'

The letter was despatched, received at Greenwich, read and answered. 'The Astronomer Royal is at present away but I feel sure he would grant your request. I should be obliged if you would ask for me on the first occasion (next Tuesday morning) that I may explain to the porter.' The chief assistant, Herbert Turner, had written the response: Ruth was in. And by now, the Belvilles were not the only women passing through the hallowed gates of the Royal Observatory to work.

When the Observatory was founded in 1675, the first Astronomer Royal, John Flamsteed, had little help. But two centuries later, William Christie led a team of some fifty staff. The full complement included eleven assistants (each managing a specific section of work), a clerk, and a large number of 'computers' – people who helped the assistants carry out the unending routine calculations involved in astronomical work. Some of these computers were on the salaried staff, who might stay for many years, working their way up to assistant level. Others were 'supernumerary' – young boys, 13 or 14 years old, fresh out of school and good with figures, wanting to stay only a year or two while they built up experience. They were hired and fired by the Astronomer Royal himself as required. Until 1890, the Observatory staff had been exclusively male, although earlier in the institution's history, a very small number of women had performed off-site computing for the *Nautical Almanac*, paid per job, never salaried, and rarely if ever coming to Greenwich.

But in 1890, William Christie embarked on a trial that was to last for just five years, an experiment which meant that Maria and Ruth would not have been the only women on-site when they came every week to check the time. The Royal Observatory was booming, new buildings and instruments were being built, and as part of his expansion plans, Christie hired the first ever 'lady computers'. These were university-educated women, not doing the mundane and tedious numbers work that the boy computers did, but proper astronomical observation. Their job title, though, reflected the prejudices of the times: Christie had to be able to hire them from his computing fund, so that he would not need to seek permission from a very conservative Civil Service. The down-side – and it was a huge down-side for the women – was that they were paid a pittance compared to men doing similar work. Their average salary was something like one-sixth that of a male assistant. They didn't even show up in the Observatory's annual reports – a symptom of the fact their presence had to be kept quiet. Christie was very open about this restriction, which was made clear in discussions with potential new applicants. 'There are at present four ladies employed as computers at this observatory, these being the only appointments in the hands of the Astronomer Royal, and therefore the only ones in which it is possible to employ ladies without raising the question of their recognition by the Civil Service Commissioners'.

To give Christie his due, he was working in a world that was very much against women taking on positions such as this with anything like equality. In appointing the women as computers he was using the only route open to him and there is evidence he was both influenced by, and sought advice from, the campaign for the advancement for women that was in progress at that time. In fact, Christie's scheme was warmly welcomed by such activists. How else would women get their feet in the door of scientific employment? In the end, though, his enlightened approach failed. Only four women ever took part in the scheme and they had all left by 1895, either to marry, or to seek more remunerative employment elsewhere. The money just wasn't good enough, and the programme was over just five years after it had begun. It was to be many years and a world war later before women again joined the payroll of the Royal Observatory.

Outside the short-lived programme of Christie's lady computers, women had to take their own chances working with the men at Greenwich. Maria Belville had been lucky to have a daughter who could take over her business so readily, but of course Ruth had been trained up for the work since she was a small child. One family member later recalled that Ruth 'was treated as a grown-up before she was six years old', always by her mother's side, and she was certainly an observant girl. She remembered details of her time trips, recalling, for instance, that some watchmakers could not afford her mother's annual subscription and had to resort to hand-me-downs. Ruth explained, 'I myself have a sort of recollection of a firm in Clerkenwell where I went with my mother when I was a small child . . . after she had checked the regulators by the chronometer . . . we passed three or four people going in to the shop, chronometer in hand, and my mother telling me that these people were working chronometer-makers who paid a small fee to the big firm so that they might obtain the time second hand!'

A second-hand second hand. This was exactly how all time supply chains worked. A primary standard sits at the top – the stars passing over the Royal Observatory every night. A clock is set by the stars. Another clock is set by that clock, which sends out time signals to intermediate time stations – post offices, for instance. Postmasters use the incoming time signals to set their office clocks; people visit to check their watches against the post-office clocks, and so on down the line. Time trickles downwards from Observatory Hill, and not just to post offices. Time signals, directly or indirectly connected to the master clock at Greenwich, could be found in all sorts of unlikely places around Victorian London and other big cities. There was the telegraph company's time ball opposite Charing Cross railway station. A miniature version could be seen dropping every hour in the window of Gledhill-Brook, a London clockmaker, and jewellers' shops proudly displayed electric telegraph instruments with big red needles that swung left and right when the hourly time signal came through.

It was inside this complex network of electric technology and commerce that Maria and Ruth worked their own trade. Time was money, and the Greenwich time ladies knew very well that these new electric competitors would try to muscle in on their territory. Their customers were loyal, but not stupid. They needed to be reminded that the Belville service suited their needs and was the best option available. The Belvilles thought a little public relations activity might help, and the new pictorial press was one way they could keep their stock high. The occasion of the handover from mother to daughter was a perfect opportunity for some 'advertorial' – but if you dance with the dailies, you might get burned.

Four months after Ruth received formal permission from the Observatory to take over the business, the two women were involved in some coverage in the *Daily Graphic*, a pictorial daily newspaper launched two years previously. It featured a prominent photograph of Maria Belville bringing 'the time' to a customer's premises, probably that of the John Walker firm which had a shop in Regent Street. A good-quality pendulum clock is in the background, the sort of device that all good watch- and clockmakers would possess, and on the table beside Maria are two marine chronometers, the beating heart of marine navigation and a symbol of accuracy and prestige.

The portrait was taken by Samuel Walker, a photographer living near London's Paddington station and also working in Regent Street, though not apparently related to John Walker. The article, entitled 'Greenwich Mean Time', explained the scene. 'The accompanying portrait is that of an unknown but not unimportant personage who has for the past forty years enabled Mr John Walker, a railway clock-maker, to supply the principal London railways with Greenwich mean time.' Maria's work was put in the context of the Astronomers Royal and John Belville's role as the Royal Observatory's original time gentleman. Then came the sting in the tail. 'Although mean time is supposed to be transmitted by electric current, it is a well-known fact that the correctness of the current is not to be relied on.' Anyone operating an electric time service in London that day in 1892 must have smarted at this comment. The article concluded by establishing the personal credentials of the Greenwich time lady and stressing the continuity of service stretching decades: '. . . comes of a very good Norfolk family . . . accomplished linguist and mathematician . . . at the advanced age of eighty-one Mrs Belleville is compelled to hand the duty over to her daughter, who thus continues the important work which has been carried on by the Belleville family for over sixty years.'

Putting aside the misspelling of Maria's surname, which often happened to the family, this is an example of how trust is built, a classic case of working out how customers *think*, what they value and are prepared to pay for. Good pedigree and a long lineage, continuity of service, the personal touch and a friendly face. This article was all about winning hearts and minds in the face of stiff competition from the *arriviste* electric time suppliers, by casting doubt on the reliability of their product. The trouble was, there were only two electric time suppliers in London as the nineteenth century drew to a close, and with comments like this attributed to them, the Bellevilles were not going to win any friends.

The first was the Royal Observatory, which had been sending out electric time signals since Charles Shepherd's electric clock had been installed in 1852. The Observatory did not have private clients itself, only the Post Office, a handful of direct lines to Admiralty coastal signal stations and one to the great clock at Westminster (Big Ben). It was the Post Office that distributed this form of Greenwich time onwards to its branch offices, the railway system and to a few private customers. It was not huge competition for the Belvilles, whose clients tended to be individual businesses, but the comments in the newspaper did not go down at all well with the officials at Greenwich. Herbert Turner wrote immediately to the editor of the *Daily Graphic* to express his concerns. 'I am sure Mrs Belleville, of whom you gave this morning an excellent portrait, would be the first to regret the misstatements made by your correspondent . . . the correctness of the Greenwich time signal, as distributed telegraphically by the Post Office may be thoroughly relied on, although the responsibility of the Astronomer Royal extends no further than the transmission of this signal hourly to the Post Office.' The Royal Observatory's time service was being called into question, and Turner could not let readers think that Maria Belville had the upper hand. He went on, 'Her present usefulness consists, I believe, in supplying the approximate time to those who find the Post Office charges too high.' To conclude, Turner could not be seen to be denigrating Maria's work too much, so he explained that 'I do not think that any member of the Staff of this Observatory has any but the kindliest feelings towards Mrs Belleville: but the Astronomer Royal could not allow to pass unnoticed the statements of your correspondent.' His letter was published in full the following day.

This wasn't a good start for Ruth Belville. In 1856, Maria had incurred the wrath of George Airy a few months after taking over the business when he thought she had used a key inappropriately. Had Ruth just angered William Christie a few months after *she* had taken over, by criticising the very accuracy of his work in the national press? Her mother dashed off a letter to the Observatory referring to the correspondence in the *Daily Graphic*. 'I would beg you to understand that any mis-statement that may have arisen therein has in no way emanated from my daughter or myself. Thanking you for the kindness and attention I have received at the Royal Observatory for so many years.' This was enough to pacify the men at Greenwich but, as a matter of fact, Turner's own letter had been self-serving and, indeed, a little unfair. The evidence shows that the Greenwich telegraph time service, as delivered through the Post Office, was certainly *not* always reliable. There is an Observatory file an inch thick entitled 'Post Office Time Signals – Complaints and Failures', with corresponding files in the archives of the Post Office telegraph department. The signals were wrong or absent often enough for a great deal of mistrust to build up. This was by no means a straightforward system and the split between the time men and the wiremen, as Turner highlighted, was far from satisfactory. When things went wrong, and they often did, each side blamed the other.

Moreover, Maria Belville supplied much more than just an 'approximate' time, as Turner put it. Her pocket chronometer, looked after by the Observatory's own staff, kept time to a tenth of a second, and she received a certificate to that effect each time it was checked. Assuming that Maria completed her rounds the same day the check was made, the time she was peddling up in London each week was very much better than approximate – it was excellent and, what was more, consistent. Customers not only received the right time each week but also knew how the watch had performed over the previous week; building up a years-long record of this performance would have given an excellent picture of its stability and how trust-worthy it was.

Even the press reports picked up on this important technical point. A few years later, in 1913, the Post Office was about to set up a cheaper version of its electrical time service, known as 'synchronisa-tion'. A newspaper article noted that this, 'it might be thought, would risk the business of the lady who calls at the Observatory once a week for the right time and then carries it around to her clients. But appar-ently there is no danger of this. Miss Belleville . . . cuts time down to finer distinctions than any synchronised clock can aspire to.' The Belville service simply worked, which was more than could be said of the telegraph time system. It was this that was approximate, at least some of the time, and it was not knowing when it was playing up that caused problems for electric time renters. The complaints file makes very clear that the signal sometimes went wrong by up to a second or two owing to technical difficulties with the complex equipment, and sometimes it would not be received at all for days on end.

Was this such a good service? By contrast, while the Belvilles' dis-tribution technology – a human being carrying a watch in a handbag – seemed antiquated compared to the copper wires of the telegraphs, the timekeeping technology, which was a good stable chronometer set to an excellent time standard and certificated to a tenth of a second, was excellent, and better than time-by-wire. All things considered, the telegraph time system was fine for some customers, but the specialists needed something better, which they could trust, and that was what the Belvilles provided.

Finally, there was the question of cost. Turner was right to say that Maria Belville charged less – much less – than the Post Office, although the evidence here is scant. It is thought that she charged something like £4 per year, compared to about £15 for the full Post Office subscription. But here's the rub. What if there was another electric time service, derived from Greenwich, reliable and costing less than half that of the Belvilles? What if this new service also provided customers with clocks automatically corrected every hour by the time signal, with no human attention needed? Surely, if Turner's argument held, that would sound a death-knell for the Belvilles' 'approximate' service that was only maintained because it was cheap. Surely everyone would have switched over to this electric alternative?

Well, there was such a service and many people did make the switch, but not all. At the time of Maria's retirement there were still at least forty customers on the Belville books – enough to keep them afloat and enough to call Turner's quick dismissal into question. If the Post Office and the Royal Observatory calmed down soon after the *Daily Graphic* article appeared, there was another group of men keen to poach customers if they could. The Belvilles' real competitor was not the Post Office, but a commercial company called the Standard Time Company Ltd (STC) which ran the half-price hourly service, and STC was hungry for new customers.

STC was formed as a spin-off from a well-known chronometer-making firm called Barraud & Lund. John Lund had invented a clock-synchronizing device that could be fixed to an existing wind-up clock and connected to a central control station which sent out hourly electrical time signals by wire. On receipt of each hourly signal, the synchronizers would automatically set their clocks right using a little electromagnet fixed inside the mechanism. It was like having someone visit you every hour with a perfectly accurate watch, checking each clock on your premises, and correcting any found to be wrong.

Barraud & Lund started an experimental service in the 1870s from their premises in Cornhill, London. There was a lot that could go wrong but Lund had thought of solutions to the problems. The master pendulum clock – the timekeeper that sent out the electric pulses – could get out of step or stop altogether, so Lund added a back-up clock with an automatic switchover in case of fault. The incoming daily signal from Greenwich, which was used to keep their own clocks on time, might be missed, so Lund invented a device to record its arrival automatically. Extensive fault-finding devices populated the entire network, even down to a special machine that would prevent errors caused by random electric pulses when the overhead telegraph lines in London streets blew together in the wind. In an 1876 description of the system, Lund concluded, 'Little do the thousands of persons who daily test their watches by the deflecting needle in the chronometer-maker's shop, or office, know the enormous amount of brain power and experiment that has been bestowed to obtain so apparently simple looking a conclusion.'

After a decade of experimentation and the exercise of a whole lot of brain power, Lund was ready to make a break from the parent firm. In 1882 he set up the Standard Time & Telephone Company, keen to exploit the new world of telephony as well as operating his time signals. Both involved wire networks and subscriber-operated devices, usually leased, so the match was good. The new separate company took up premises in nearby Queen Victoria Street and began sending out its hourly signals to a growing subscriber list that eventually reached several hundred. The telephone side of the business never really took off in the face of a dominant monopoly provider, the Post Office, which could prevent any rival from operating on its turf. But the Post Office, despite running its own time-supply business, seemed content to let the new firm trade its time signals, even using Post Office telegraph lines to do so. The relationship was complex and shifted over time but, essentially, the Post Office felt it had enough on its plate operating telegraphs and telephones without worrying too much about time signals.

In 1886, the firm floated on the stock market under a new simpler name, as the Standard Time Company. By now it had over 300 customers across London, including banks, city firms, railways, shops, newspaper offices and a great many public houses. In fact, a quarter of STC's business in 1886 came from London pubs and other licensed premises. In the 1870s, new liquor licensing legislation had been passed that set strict drinking hours. Any licensee caught selling alcohol after hours was liable to lose his licence, so the stakes were high. What better defence against prosecution than a big clock on the wall of the pub, which the landlord paid to have set every hour to Greenwich time down electric wires? For the first time, publicans *needed* to know the accurate time of day: their business depended on it. And it was not just the pubs. Factories were faced with new working-time legislation that meant they needed clocks set to a common standard. Railways were operating in greater number and farther than ever before; their timetables and safety systems relied on a common time being available across the network. Everywhere, time was in the air. In 1880, as we have seen, the Time Act was passed which established Greenwich time as Britain's standard. Then, in 1884, just eight years before Maria Belville retired and handed over the Arnold watch to her daughter, the world came together at an international conference in Washington DC to discuss a time system for the whole globe.

The 1884 meridian conference discussed many things relating to time, but two key aspects were considered that affected everybody in the business of Greenwich time supply: the creation of a prime meridian and of universal time. We know the time system now as 'time zones', although their adoption has taken far longer than most historians have traditionally admitted. Even today we do not live in a world of twenty-four neat, simple one-hour time zones with straight sides. Politics gets involved, and geography, and the world time maps of today are a riot of jagged-edged zones, some huge and some tiny, and several are some fraction of an hour different from their neighbours. But the crucial outcome of the 1884 meridian conference was the selection of one place on Earth that would act as a common reference point, a single baseline, a *prime* meridian. It would be the point of origin on Earth for both time and maps.

The choice was Greenwich, and specifically the centre of George Airy's huge telescope in what is now the Meridian Building of the Royal Observatory, which from 1884 became the reference point for universal time. It was the ultimate source of time for Maria and Ruth Belville, the Post Office, the Standard Time Company and everybody else in London, around the country and now across the globe, who wanted an answer to a simple question: what time is it? It was no longer so much about knowing the real time of day wherever you happened to be; rather, it was about knowing a common, standard time that everyone could agree on, whichever timetable you consulted, whichever pub you chose to drink in. The time legislation and conferences of the 1880s had been designed to connect people up, to synchronize activity, to urge everyone to march to the same beat. Of course, it was nothing new; what changed, perhaps, is the scale over which that standardization occurred, from small towns and villages to whole cities, then whole countries, and then large parts of continents. This change of scale was prompted in part by new transportation technologies (the railway, the steamship, the automobile), and tele-communication (the electric telegraph, telephone, and later the wireless). People could move and communicate with each other farther and faster. Continents, previously separated by weeks of travel on board sailing ships, were now connected by electric cables that promised communication in hours or even minutes.

The world was coming together and standardized time was the operating system that kept everything in step. It makes sense, then, that a firm called the Standard Time Company set up in business in the 1880s, and it follows that long-standing time suppliers like Maria and Ruth Belville would be nervous of new competitors. Their world was getting more complicated as every year went by. But as the new century dawned, Ruth had to keep visiting Greenwich, taking the tram from Charlton and climbing Observatory Hill to collect the time on her Arnold pocket chronometer every week before making her way to London.

New attention was also being paid to the Royal Observatory and Greenwich Mean Time. It was becoming an icon, a symbol of science, of imperialism and of progress. Ten years after the meridian conference at Washington, a new Chief Assistant joined William Christie at the Observatory. This was Frank Dyson, who was to succeed Christie as Astronomer Royal in 1910 but who in February 1894 had just arrived at Greenwich and was staying in nearby Blackheath as he found his feet. His arrival, however, coincided with an event which truly catapulted the Royal Observatory into the limelight.

It was the afternoon of Thursday 15 February 1894. Martial Bourdin, a 26-year-old Frenchman, left his lodgings in Fitzroy Street, west of London's Tottenham Court Road. He dined, then walked to the tram terminus at Westminster Bridge, near the Houses of

Parliament. There, he climbed aboard one of the horse-drawn trams which ran every few minutes, buying a through ticket all the way to the end of the line at East Greenwich. At Greenwich, Bourdin seemed to be in a hurry and was agitated, as if he had gone further than he intended or was late for a meeting. He called out to staff at the tram stop, 'can you direct me to Greenwich Park?' The tram conductor was standing next to his colleague, the timekeeper-on-duty, who had noted that tram number 379, which was due at 4.16 p.m., was about a minute-and-a-half late (timekeeping is important in Greenwich). One or other of the men directed Bourdin into the Park. This was all reported to police after the incident.

A little later, two of the Royal Observatory assistants – William Thackeray and Henry Hollis – were at work in one of the computing rooms, at the top of the hill overlooking the northern reaches of Greenwich Park. Reports put this at about 4.45 p.m. Also on-site at the Observatory besides the assistants was the gate porter, William McManus. This was the man who opened the gate to Ruth Belville every week so that she could check her pocket watch. He was an old sailor and he had seen and heard plenty in his time. But who was expecting anything to shatter the calm of the Observatory that Thursday afternoon in February 1894?

All three men heard the explosion. In fact, it could be heard across Greenwich. At a coroner's inquest later that month, some witnesses described the sound as sharp and clear, but McManus characterized it as muffled, not at all like the sound of a cannon firing – a sound the ex-sailor knew well. They hurried towards windows, railings and other vantage points. Smoke was seen rising from the trees near the Observatory's front terrace. Had somebody been shot? McManus ran towards the explosion site, seeing Thackeray and Hollis do likewise.

First on the scene, though, were two local schoolboys, George Frost and Thomas Winter, both pupils at the Roan Boys' School, East Street (now Feathers Place), Greenwich – a building just a few feet from the site of John Belville's first house after leaving his lodgings at the Royal Observatory in November 1822. At the site of the explosion they found a man, later identified as Martial Bourdin, kneeling on the path by the railings, perfectly still. His head was bowed. The Park keeper on duty that afternoon, Patrick Sullivan, was next to arrive, followed by McManus, Thackeray and Hollis. At that moment, Bourdin was seen to sink to the ground. Sullivan lifted him gently upright and spoke to him. 'What have you done?' No answer. The witnesses had by now discovered that the man's left hand had been blown off; sinews and tendons were hanging down out of the bloody stump. He had a massive wound in his stomach, out of which some of his intestines were spilling, and he had a hole under his right shoulder blade with bone protruding. One of the boys was sent to the nearby Dreadnought Seamen's Hospital to fetch a doctor. Bourdin murmured that he wanted to be taken home. Dr Willes, soon on the scene, sent one of the Observatory men for brandy, which was used to moisten Bourdin's lips. His eyes opened at that point. The gathered party took Bourdin to the hospital where he died twenty-five minutes later from shock and loss of blood. He never said what had happened.

McManus, Thackeray and Hollis returned to the Observatory and performed a search of the area between its buildings and the path nearby, where Bourdin had been found. It was a gruesome experience. They found many fragments of his hand, including a two-inch piece of blackened finger-bone. Blood and clothing fragments littered the scene and, the following day, detectives discovered pieces of tendon wrapped around nearby railings, and two knuckle-joints from

Bourdin's left thumb. Martial Bourdin, an anarchist terrorist, had accidentally blown himself up with a bomb right outside the Observatory buildings.

The press reports the next day were almost hysterical but an editorial in *The Times* two days after the explosion took a more reflective view of events. 'It seems reasonable to suppose that he stumbled and came into unexpected contact with the Earth, with the result of being in the most literal sense "hoist with his own petard." What he was doing with his petard in Greenwich Park must probably remain to some extent a matter of conjecture.' The Park was packed with sightseers for days afterwards, thronging the path where Bourdin had come to grief. Why had he carried a bomb to the Royal Observatory? Was the home of Greenwich time his intended target, or was he simply seeking a quiet place to get rid of explosives in light of increased police surveillance? We will never know the bomber's motives for sure. Perhaps he was simply passing through, hastening to some other, busier target. But the path he was on led only to the Observatory. It seems much more likely that the iconic nature of the institution, atop its hill in Greenwich, now made it a target for attention-seeking political terrorism. This was the home of international timekeeping, a single reference point for all charts, maps and time zones. Since 1884, the line passing through the Airy telescope, through the courtyard and into Greenwich Park, had become the world's prime meridian, the line that bisected the western hemisphere and the eastern. All time signals came from here; Greenwich Mean Time originated here. The dial in the wall outside McManus's gate-lodge was the symbolic first clock to show GMT to the public. The bomb was quite small, not big enough to do real damage to a solid building. But perhaps the bomber had a smaller, more vulnerable target in mind. Conjecture, certainly, but perhaps Martial Bourdin was visiting the Observatory that afternoon to blow up the gate clock. Perhaps he wanted, publicly and violently, to stop time.

What is certain is the extent of public interest in this act of terrorism on British soil so close to the capital. The newspapers carried reports for weeks and months afterwards, and the episode became the focus for wider expressions of unease at the threats in our own land. Whether intended for the Observatory or not, the explosion forced itself into the wider cultural consciousness. Joseph Conrad based his classic and influential 1907 novel *The Secret Agent* on the Greenwich bomb attack. This inspired Alfred Hitchcock in his 1936 film *Sabotage*, and in the 1990s, a series of explosions caused by Theodore Kaczynski (known as the 'Unabomber') over an almost twenty-year period were said by the FBI to have been inspired by Conrad's work. On 15 February 1894, the clocks almost stopped. Ruth Belville would celebrate her fortieth birthday a fortnight later. What a celebration, picking her way through the detectives and journalists and sightseers that newly filled the narrow path to the Royal Observatory, ringing the bell and asking William McManus if she could come in. He had kept the gate for seven years before the bombing, and stayed for a further nine. What was the conversation like that month between the two Greenwich veterans over a cup of tea, the peace and quiet of their Royal Observatory having been shattered by Martial Bourdin?

Time passed quickly for the two Belvilles as the 1890s continued. In 1896 (a year in which STC, incidentally, was financially on its knees and in temporary receivership), the women moved a little further up Elliscombe Road to number 63, and then again the following year, two doors further to number 67. In 1899, they were on the move once more, to what was to be their last Charlton residence, around the corner to 25 Wellington Road (now Wellington Gardens). This was Maria's final home. On 29 December 1899, the first Greenwich time lady passed away, aged about 88. She died at home from simple old age, with Ruth by her bedside, and was buried five days later in nearby Charlton Cemetery in an unmarked, grass-covered grave.

What a life. Maria Belville had worked on important meteorological observations, taught French to astronomical assistants and the wife of a French admiral. She had buried a husband when only in her forties, run a boarding school, negotiated with one of the most powerful scientists in the land and had never given up when the odds were stacked against her. She had brought up a little girl alone, and for forty years had run a time-supply business that helped keep a thriving Victorian London on time. Maria Belville had lived through a period of extraordinary change, her span almost matching that of the reigning monarch, Queen Victoria. But just two days before the numbers on the calendar changed to 1900, Maria's time had ended.

Ruth Belville, now 45, was on her own and stayed at Wellington Road until 1907. She may have given up teaching French and music, as the 1901 census simply says she was 'living on own means' – including the income from four boarders, all engineers and electrical apprentices in their twenties. What irony. Electric time signals and electric clocks were (people thought) threatening the existence of the Belville time service; and yet here was Ruth providing lodgings for four new-breed electricians. Every week, Ruth attended the Royal Observatory, and every week she continued to sell the time. After the anarchist bomb, she must have thought nothing else could rock her world. But she was wrong. A group of electrical men, like her own lodgers, had other ideas.

Chapter 4

LYING CLOCKS

{ January–March 1908 }

Astronomer Royal William Christie at work, about 1890.

43 ST LUKE'S ROAD, MAIDENHEAD, FRIDAY 6 MARCH 1908. It was evening, and Ruth Belville was at home. This was now in Maidenhead, a small town to the west of London with a fast railway line into Isambard Brunel's magnificent Paddington station, and Ruth had moved there the previous year, leaving behind Wellington Road, Charlton, and its easy access to Greenwich. Her new home was a small ivy-covered brick semi-detached house in St Luke's Road, and this spring evening in 1908 there was a knock at the door, which Ruth's servant answered. Two gentlemen wished to see Miss Belville on urgent business, said the maid. Standing on the step, to Ruth's surprise, were newspaper reporters from London's *Daily News* and the *Evening News*, who asked to come in to discuss her time-supply business. They had heard about her in a lecture, they said. She asked them to hold off until Monday. She wished to speak first to the Astronomer Royal, William Christie, at the Observatory. She promised she would then call at their offices

to give them the information they sought. They brandished reports they had pieced together. You've got it all wrong, she cried. I can correct these, she said, once I've been to the Observatory on Monday. But just hold off, please. Not a chance, they said. It will be in the papers with or without any information on your part, they said. We go to press tonight, like it or leave it.

Ruth panicked. She quickly went through the written notes they had cobbled together on the train out to see her, made basic corrections to facts and showed them accuracy certificates for her watch. But really, she could not comment on her relationship with the Observatory until she spoke with them; could this story not wait? The reporters left, shaking their heads and Ruth closed the door behind them. The next day, she woke to find the newspapers full of reports about her time-supply work. Horrified at this intrusion into her quiet life, she went to her writing table and started to pen a letter to William Christie at Greenwich. Over the weekend, more pressmen arrived, national and local, all eager to uncover details of the 'lady who conveys the time', and this was not just a provincial story: newspapers as big as the *Daily Express* sent men from London in search of a scoop. The papers that could not send representatives cabled urgent telegram messages to the Greenwich time lady, although she refused to answer them. What on earth had been said in the lecture?

The talk had taken place three days previously, on Wednesday 4 March, at 6.30 p.m. when sixty-three members of the United Wards Club, plus a sprinkling of interested outsiders, gathered in meeting room 35 of the Cannon Street Hotel. This was a 'high Victorian jumble' of a building on the frontage of the Cannon Street railway terminus, and the location, as it happens, of the formation of the Communist Party of Great Britain twelve years later. The United Wards Club was a talking-shop for public, civic and guild affairs affecting the City of London. Its membership was City businessmen, municipal officials, lawyers and policy-makers. Work had finished for the day and the president of the club, Mr Huxtable, called the assembly to order. The usual committee business was transacted quickly – reading of minutes, election of new members – and then the president called upon the guest speaker for the evening. Mr St John Winne, a fellow of the Royal Geographical Society, proceeded to deliver his lecture, 'The Time of a Great City.' It changed Ruth Belville's life.

Who was this man? His history is somewhat elusive but the facts are that St Andrew St John Winne was born in 1862 and trained as a solicitor's clerk before becoming a company promoter. He helped companies raise money and he helped them make money, and to do this he had to promote his target companies so that other people would want to invest their money into them, and therefore, to a degree, into his own pocket (through fees and commissions). Company promotion was often more for the benefit of the promoter than the company. This was all permissible to the Victorians. Financial rules and regulations were not what they are today. But in Mr Winne's case it seems that some of the companies he was involved with were, even then, a little colourful. At one stage he worked as a secretary for a man called Horatio Bottomley – in fact the two were friends – and Bottomley's way of doing business was not exactly squeaky-clean. Or, as his biographer put it, Bottomley was a 'journalist and swindler' partial to a succession of 'get-rich quick schemes', a 'fraudster' with 'latent megalomania' and 'overwhelming egotism'.

Winne should probably not be tarred too much with Bottomley's brush but, after working with him on several projects in the 1880s, Winne was later involved with the Standard Time Company (STC), Ruth's rival London time-supplier. Technically STC had a sophisticated system, but the company always had trouble making any money and in the early 1900s it was floundering. In 1905, one of its directors, Noel Francis Nalder, resigned and was replaced on the board by St John Winne, who also bought a lot of shares. Winne was a money man, whereas Nalder was an electrical engineer, and Winne arrived to find a company with a sound technical infrastructure but serious cash-flow problems and a stagnant market that didn't seem to be growing. He needed to do what he was good at: to promote the company's operations so that it could make money. He needed a plan, and part of it was going to involve Ruth Belville.

Here in a nutshell is Winne's scheme. It was the middle of the Edwardian period. Business was booming, the world was going modern and in London and cities around the country, there were countless public clocks in the streets and on buildings that told passers-by the time. The problem was, they didn't all tell the same time. What if local laws were passed to oblige anyone with a public clock on their premises to keep it exactly on standard, legal time – Greenwich time, all marching to the beat of the clocks at the Royal Observatory in Greenwich. If local authorities passed a law like that, what would public clock owners need? They'd need the services of a clock synchronizing company, like STC – which happened to be pretty much the only company in Edwardian England to operate an automatic clock synchronization network since at that stage the Post Office only supplied time signals, not automatic clock correction services. This was Winne's plan. If he could convince local government to legislate in favour of clock synchronization, his own company and shareholders could reap the rewards. Of course, other companies might set up rival services, but it was technology-intensive – lots of wires, clocks, batteries and relay switches – and STC was already the leader in this industry.

Pushing for this sort of legislation was not new, at least not in London, which was the main focus of STC's attention, given the capital's high concentration of financial and business concerns. In 1904, the London County Council (LCC) and the Corporation of the City of London had adopted resolutions to the effect that all new public clocks had to be synchronized (not all the existing ones). This prompted a storm of protest in the trade press, particularly the *Horological Journal*, the monthly magazine for professional clockmakers. The old guard hated the new-fangled idea of synchronization, firstly because they made a lot of money on weekly winding and setting rounds of the mechanical clocks, and secondly because they felt that the very need for synchronization suggested their clocks couldn't keep time in the first place. Really, it was mostly natural resistance to change but the arguments they put forward centred on accusations of corruption between the LCC and STC, suggesting that 'the Municipality took orders from the S. T. Co.' But despite the London resolution seeming to favour STC, there is no evidence of the company deriving any real benefit from it through any dramatic upturn in its fortunes. A combination of the rule only applying to new clocks (of which there were very few), and feeble enforcement, meant that the 1904 LCC regulations had little effect, and the mechanical clock-makers had breathed a sigh of relief. But Winne's arrival at STC in

1905 brought new business approaches and, this time round, STC wanted legislation that would really make a difference.

January 1908 saw Winne's plan swing into action, and it became known as the 'lying clocks' saga. We need to examine this episode, because it is the background to Winne's lecture at the United Wards Club in March. That speech did not stand alone, but was the conclusion of three months of politicking. The propaganda campaign was launched on STC's behalf by Sir John Cockburn, vice-chairman and founding member of the British Science Guild, who sent a letter to *The Times* on 8 January 1908, given the heading 'lying clocks' by the newspaper's sub-editor. 'Surely', wrote Cockburn, 'there should be some censorship as to the time kept by clocks exposed to public view in the streets of London. It is not unusual within a hundred yards to find clocks three or four minutes at variance with each other.' He then blamed this lamentable state of affairs on human behaviour: 'Highly desirable as individualism is in many respects, it is out of place in horology. A lying timekeeper is an abomination, and should not be tolerated.' Then came Cockburn's clinching argument. 'A by-law might well be framed requiring clocks in public places to be synchronized with standard time; the penalty for repeated disregard to be the removal of the offending dial.'

The following day, *The Times* ran a leader article commenting on Cockburn's suggestions, warmly endorsing his sentiments, though not his solution. 'Perhaps the remedy rather lies in co-operation than in compulsion,' the editorial suggested, proposing instead that every municipal authority should first set up its own synchronized clocks, 'to show how to do it better before it proceeds to punitive measures'. This long editorial, in a prominent national newspaper, then put in a plug for STC's technology and kicked the traditional clockmakers in the teeth. 'The mechanical synchronization of public clocks is an impossibility and their synchronization by individual agency' – this meant the weekly winding and setting rounds of the old-guard clockmakers – 'a tiresome and by no means easy job.' The traditional trade was now completely wound up, but the article went on to twist the knife: 'it is comparatively a simple matter to synchronize any number of clocks by means of electricity, and we conceive that the time is not very far off when a town or city of any note or repute which is not provided in every important thoroughfare with public clocks electrically synchronized will be regarded as extremely antiquated and unprogressive in its municipal equipment.' The antiquated clockmakers were horrified.

Two days after John Cockburn's salvo, *The Times* ran a correspondence column filled with responses. First was a letter from STC's secretary, Edward Newitt, keen to capitalize on the previous day's editorial. He told readers that STC had been running an electric clock-synchronizing network for thirty years. It wasn't that the *means* to synchronize didn't exist, he exclaimed, it was all because of 'the apathy displayed by the Government, the London County Council, the City Corporation, and the public'. Newitt was quick to focus his attention. The LCC and the City Corporation were first in line, but it was the clockmakers who corrected clocks each week on their winding rounds who were holding back the tide of progress. For a start, he noted, they got their time from STC in the first place, which meant STC's subscribers were in effect subsidizing non-subscribers, keeping prices high. But the real problem was the clockmakers' technical conservatism that associated synchronization (hourly, rather than weekly) with imperfection in mechanism – *their* mechanism, the wheels and pinions *they* made – so they would rather let errors accumulate for a week and then set everything right when they came round to wind the clocks than have the hands tweaked every hour by a little electromagnet.

The next letter was from a Wimbledon writer, H. Berthoud, who compared the deplorable British situation with that in the capitals of Europe where all boulevards and main streets were stuffed with electric clock systems all telling the same time. 'Why cannot London do the same?' He wanted to see synchronized clocks at Piccadilly Circus, at Charing Cross, at every crossroads. 'Surely it is time', he observed, 'that the nation which has invented the proverb "Time is money" should not stay behind every other civilized country in this respect.' Putting aside the fact that 'time is money' was invented by an American, Benjamin Franklin, and that Berthoud's list of civilized countries consisted solely of Britain, France, Belgium, Germany and Austria, his letter was ammunition for STC's call for city-wide action to synchronize the public clocks. The next writer, Robert Orr, went even further. He mentioned STC by name (although he was at pains to point out he held no brief for them), and then described the 'criminal indifference' shown in English public life, with Londoners 'fatuously and impotently pottering about with innumerable "lying clocks"'. He concluded that the reason London didn't have a network of synchronized clocks was because of the 'disheartening indifference and crass stupidity of the public, led by stupid municipal and other governing bodies, who prate about practical work, but are incapable of appreciating the depth of the meaning of the English adage "Time is money"'. After Berthoud and Orr's letters came a shameless plug from a Holborn department store, Thomas Wallis & Co., whose director wrote to tell readers that the clock installed on their shopfront was corrected by the Standard Time Company every hour (he even sent the *Times* editor a leaflet for STC) and 'the public who consult our clock at Holborn-circus will always be "on time".' Finally, E. Tillett wrote to inform everyone that the two adjacent clocks at London Bridge railway station, owned respectively by the London, Brighton and South Coast Railway and the South-Eastern and Chatham Railway, had been different by several minutes for some time past. Tillett shouldn't have been surprised: as the two companies were fierce rivals, perhaps they just wanted each other's customers to miss their trains.

STC's plan was going well, but in the course of all this rhetoric, it had sharply criticised two main groups: the municipal bodies themselves and the traditional clockmakers. Firstly, let us hear from the one of the 'stupid' municipal bodies in question: the City of London Corporation, charged with government of London's 'square mile'. It was now 11 January, three days after Cockburn's original letter, and Deputy Alderman Millar Wilkinson wrote to *The Times* to tell readers that he had been responsible for the Corporation's 1904 order stating that 'from that date all outside clocks in the City are to be synchronized. This order has been carried into effect.' STC's Edward Newitt was quick to fire off a crisp reply. 'It is just possible that it may lead your readers to suppose that all City public clocks are synchronized,' he began, tartly. Perhaps Millar Wilkinson meant to state that from 1904, all 'new' clocks – of which there were two – were to be synchronized, he proposed, because if he was suggesting that the other sixty-two public City clocks, which already existed in 1904, were also now synchronized thanks to his order, that was not the case. Two out of sixty-four wasn't exactly what everyone had in mind, he suggested. Millar Wilkinson was chastened and later tried to mend the City's ways, but in the meantime, the mechanical clockmakers had girded their loins and sent one of their finest gladiators to battle. It was time to stand up for the trade, and who better for the job than E. Dent & Co., which signed its letters, 'makers of the standard clock (the primary standard timekeeper of the United Kingdom) of the Royal Observatory, Greenwich, also makers of the Great Westminster Clock, "Big Ben"'. The Dent company was the biggest of the big guns and the points they made in their letter, while undoubtedly reactionary and self-serving, reveal a point of wider interest in this story.

The Dent company was represented by W. Pyall, its secretary, writing from their premises on the Strand, London. Firstly, Pyall, began, 'we have clearly demonstrated the fact of its being possible for public clocks to indicate the time to a few seconds a week without the aid of any synchronizing'. True, the clock known as 'Big Ben' was a very good timekeeper, rarely being more than a couple of seconds away from Greenwich time. But on the other hand, it is rather a special clock, and really rather *large* for the average high street. But back to those electric synchronizing devices: 'in fact, to attempt to put any such fakement to a clock with any pretensions to timekeeping is not only an insult to the maker and caretaker, but is derogatory to the best interests of the craft and science of horology'. Certainly, Dent and its peers made some amazing clocks, many of which survive and work to this day. The new-fangled electrical 'fakements' – crafty, deceitful contrivances – may have insulted the old-timer but they did seem to be doing a better job at keeping London on time than the makers and care-takers Pyall held in such high esteem. What can be done in principle and what is actually done in practice are two very different animals. Pyall's venom was now flowing freely towards the time-signal men, with their telegraph wires and their hourly corrective signals. 'Above all, to make it compulsory for all public clocks to be synchronized would be monstrous,' he blustered. 'Most of us have experienced the difficulties the Post Office have to keep private lines in order, and when the line fails the synchro-nizing fails also, leaving an erratic clock that has been kept under restraint to go its own lying way.'

Pyall here crucially reveals that the underlying infrastructure for this time-distribution network, the Post Office telegraph sys-tem, was in practice fallible. The day-to-day experience of users was not always of a reliable service. This is an important wider point in the history of technology. Maintenance and the user's perspective are far more important than many historians recog-nize, since they tend to be interested more in invention, the act of first creation. In reality, the story is always in the use of technolo-gy rather than its invention, and here was a user talking about con-tinuing problems with the Post Office system decades after it was first invented.

The correspondence rumbled on for a few more weeks, then John Cockburn sent a second letter, in early February, drawing together the strands of the argument, which essentially were these: it was important for cities to have public clocks telling the right time; London should have been leading the way, but wasn't; the mechanical clockmakers believed in perfect clocks, but that was a fallacy; and stopped clocks are worse than wrong ones, because twice a day they look nearly right, so 'they should be made to stand with hands folded to some predetermined hour so as to minimise the risk of misleading'. The propaganda campaign devised by Winne for STC was going well, but there was a fly in the ointment. For his plan to work, he had to have virtual monopoly control over the time-distribution market, because even if laws were passed obliging public clock owners to set their clocks to Greenwich time, an infrastructure-intensive and technically demanding business such as STC's would only pay if it cornered the market of all potential subscribers. The trouble was, there was a rival, which already had a subscriber book of about forty customers, some 10 per cent of the number of subscribers STC served. Forty more would certainly make a difference to STC's bottom line. Moreover, that rival business had few infrastructure overheads and it brought Greenwich time to subscribers directly from the Royal Observatory accompanied by official certification which STC did not have, although not for want of trying. That rival was Ruth Belville, the woman who took Greenwich time to her London customers every week using a pocket chronometer, as her mother had begun doing before Winne was even born. It was time, thought Winne, to get personal.

His lecture to the United Wards Club was essentially a live version of the correspondence that had been printed in *The Times* over the previous couple of months. Having followed that correspondence, it is now clear that the lecture – to an audience of London municipal officials – was STC's attempt to sell them the need for legislation obliging public clock owners to keep their clocks synchronized to Greenwich time. John Cockburn was in the audience, as was Millar Wilkinson, two men already very familiar with the lecture's contents. But listening intently at the back of the room, clutching a sheaf of notes and papers, was a new face: Daniel Buckney, managing director of E. Dent & Co., the old-school clockmaking firm whose secretary, Pyall, had stuck up so pluckily for the traditional trade in the recent newspaper correspondence. Winne got started on his lecture. 'Down to the present day', he began, 'no horologist has produced either a clock or a watch, from the finest chronometer downwards, which was free from error, and did not require more or less frequent correction.' Clocks of course needed to be set right every now and again, but the question was how it was done, who did the job, and how well they carried out the correction. Winne continued: 'Adjustment, with comparatively few exceptions, was the work of human hands, and on all sides the result was a dismal failure, and the lamentable state of affairs against which the outcry of the public was now heard.' But guess what? Apply electric synchronization to these public clocks and 'no more would be heard about "lying clocks"'. To cut a long story short, the audience was treated to a long advertisement for STC's work. The old appeal was raised for legislation to oblige all owners of public clocks – municipal or private – to keep them synchronized, and the best way forward, Winne suggested, was for the Corporation or LCC to convene a conference to discuss how to progress this great project of unifying the time of a great city. 'If such a conference were convened, he would not be guilty of egotism in suggesting that the members might usefully consult his associates.' He sat down, confident that his remarks had hit home. Then the discussion began.

Millar Wilkinson moved the vote of thanks, concluding that the Corporation was now reconsidering its position on the synchronization of all existing, not just new, public clocks in the City of London. 'They should be pioneers,' he said, and opened the meeting to the floor for comments. After such a casual dismissal of the weekly winders and setters, Daniel Buckney of Dent & Co. was soon on his feet. Brandishing official reports of the timekeeping of his firm's 'Big Ben', he pointed out that the public *did* have access to public clocks that kept good time. Smarting from the 'dismal failure' jibe, Buckney launched his own counter-attack against Winne's electric synchronizing service. 'Synchronisation was behind the times. He was a practical clockmaker and it was his aim, without saying it as an advertisement, to make clocks of as fine a quality as possible.' His trade was under attack and he pleaded for support: 'He hoped the Corporation would not be led away by the Synchronising Co. into adopting this, but support most heartily and sincerely the efforts that were made to produce correct time by private enterprise.'

The discussion descended into attacks on Winne for using the lecture merely to advertise his own company's work and his attempts to create a monopoly situation in which only STC could provide synchronizing services. Not true, he was at pains to point out. In fact, he suggested, if the laws that he advocated were passed, the business would be thrown open to competition and he was sure that Dent & Co. would be among the first to compete for the work. He was, he said, very pleased to see Buckney in the audience and while he thought some of the criticisms were a little unfair, he could afford to pass over them and ultimately hoped the meeting had been of some use. Some use to him, certainly, and Buckney too had managed to get in plenty of plugs for his own company.

But Winne had decided to open the competition a little early. There was still that fly in the ointment, that irritating little rival who was already distributing Greenwich time around London. She needed to be put in her place, and what better way to dismiss the work of the Belvilles than to offer the men in the audience a little light relief. 'It may be interesting and amusing to some of you to learn how Greenwich mean time was distributed amongst the clock and watch trade in London before the present arrangements came into vogue,' he began. 'A woman possessed of a chronometer obtained permission from the Astronomer Royal of the time to call at the Observatory and have it corrected as often as she pleased. She then made it the business of her life, until she reached a great age, to call upon her customers with the correct time, and on her retirement this useful work was, and even today is, carried on by her successor, still a female, I think.' Talk about damning with faint praise. This 'female' still had a subscriber book of forty customers, all of whom were choosing to take her weekly service in a world where STC could provide it hourly and at a cheaper rate. That was serious competition and Winne was prepared to resort to dirty tricks to win it.

What he *actually* said to this audience of municipal dignitaries and old-fashioned clockmakers, according to one report, was that 'perhaps no mere man could have got it', meaning permission from the Astronomer Royal. Was he suggesting, by any chance, that Maria Belville had used her womanly charms to get past the Greenwich gates each week? This 'amusing' little anecdote raised a few laughs among the City men, but it went further than that. It was sufficient to set the national press scrambling for the next train to Maidenhead to doorstep Ruth Belville at her little cottage that Friday evening, as they competed to uncover the story of the lady behind the watch. Thanks to St John Winne, Ruth's world was about to unravel. We get back to where we started: the journalists knocked at her door, the articles appeared in the press, and Ruth Belville, dazed, picked up her pen.

To the Astronomer Royal, Saturday 7 March 1908. 'Dear Sir, I have to regret exceedingly that my name and Chronometer work is being commented upon by some of the daily papers at present. To my great surprise on Friday evening last a representative of the Press called upon me with an extract from a lecture given in the City informing me for the first time that my work had been called into question by the lecturer.' Ruth told Astronomer Royal William Christie that the hack had already pieced a report together, 'and told me that it would be all in the evening papers with or without any information on my part.' This threat had left Ruth shaken. She continued, 'I lead so quiet a life that it is most unpleasant for me particularly in a little town like this to have so much interference in my affairs.' If Ruth thought she might be in trouble with the astronomers at Greenwich, William Christie was not in a hurry to let her off the hook. Scribbled in red script across the top of Ruth's letter, still preserved in Christie's archived papers, is his terse response: 'No answer'.

But we can cheer the steely resolve of the Maidenhead woman with the silver watch. The United Wards Club felt that Winne's lecture 'gave rise to an excellent debate and although the criticisms were somewhat hard and free, Mr Winne showed excellent temper in his replies.' But Ruth wasn't so sure Winne would have thought the debate was excellent: 'I think the Standard Time Co will not attack me again in public as the result ended in rather a heated discussion at the end of the lecture and the last thing that Mr St John Wynne wanted was to advertise [my] Chronometer at [his] Company's expense which was all the result he obtained.' Ruth may have been shaken, but she was determined to have the last laugh.

Chapter 5

ALL OVER LONDON

{ March, 1908 }

Ruth Belville in the Daily Express, *1908.*

GREENWICH PARK, MONDAY 9 MARCH 1908. Ruth Belville was standing in front of the famous 24-hour Greenwich gate clock with an attendant from the Royal Observatory. They were holding a piece of paper and a pocket watch, and the hands of the clock that had been telling Greenwich visitors the time of day since 1852 revealed that this scene was being enacted at twenty-six minutes past eleven. The purpose? An agency photographer from Fox Photos, contracted by the *Daily Express*, was taking the first known photograph of Ruth going about her business. That Monday, the *Daily Express* was running an article on her work, and they wanted to print a photograph as soon as they could get one. How convenient that Miss Belville visited Greenwich on Mondays. The next day, the paper revealed to the country the face of the woman who sells London the time.

It was all a set-up, of course, as is often the case. The real photograph of Ruth Belville having her chronometer checked every

Monday morning would show her sitting in a chair drinking a cup of tea with the gate porter while an Observatory assistant was elsewhere checking the watch against the master clock, before scribbling out a certificate recording its error that week. Business transacted, Ruth would put the watch and certificate into her handbag before leaving the Observatory's confines and heading for London. But the Observatory's white-dialled gate clock is great for press photographs, and certainly better than the porter's armchair.

The journey from Maidenhead to Greenwich that day must have seemed interminable. The previous Saturday, Ruth had written her letter to William Christie, the Astronomer Royal, but what would he say to her when she arrived? She had plenty of time to dwell on her predicament as she set out that day. First came the brisk twenty-minute walk from St Luke's Road to Maidenhead railway station. Today, the fast diesel trains can do the run to Paddington station non-stop in less than twenty-five minutes, with the more frequent stopping services taking about ten minutes longer. It was then a little slower, but not much. Let's assume Ruth took the 8.59 a.m. service and had got to Paddington at 9.37. She would have been en route for an hour already, and she then would have had to struggle across town in a journey that has hardly changed in the subsequent hundred years, and can still take a good hour-and-a-half, from Paddington platform to the Royal Observatory gate.

It might have been a struggle but Ruth had numerous travel options, even then. London's public transport network in 1908 was booming. Underground electric tube railways were being tunnelled at a rate of knots. The tramway and bus system was extensive, and larger 'cut-and-cover' railways – you dig a huge trench along the streets, build a railway in it, then cover it over again – had been encircling London's dense central area for decades. What is now called the Circle Line was built to connect the mainline railway termini into one giant loop, enabling passengers such as Ruth to get from one to the other with relative ease. To get to Greenwich, Ruth needed to make for Charing Cross station, so the Circle's tracks were one choice open to her although, then as now, it wasn't exactly fast and certainly wasn't direct – and the trains were hauled by steam locomotives. An alternative option had just opened nine months previously. Originally termed the Baker Street & Waterloo Railway, one of a clutch of cheap, fast and clean electric tube lines opening up in the Edwardian decade, the service from north-west to south-east London was swiftly dubbed the 'Bakerloo' by newspaper columnists and the name was officially adopted four months after its grand opening in March 1906. In June 1907, the line was extended west to Edgware Road, a short walk from Paddington station, which itself was connected to the line in 1913. This was the modern way to travel. Descending into the deep tunnels at Edgware Road, Ruth would have bought a cheap ticket to Charing Cross, and alighted a few minutes later, her black dress smoke-free, before heading for the suburban railway to Greenwich.

At Greenwich the walk to the Royal Observatory is long and steep. It takes about twenty minutes at normal pace but, being at the top of a hill, the final five minutes leave visitors breathless and panting as they reach the summit. The trick for making a dignified entrance is to go up around the back of the Observatory, approaching the complex from the southern, Blackheath end. It's less steep but takes longer and gives you more chance to think, if there's something on your mind. Ruth certainly had something on her mind that Monday morning in March, 1908. Her final words to William Christie, written in her apologetic letter to him the previous Saturday, were that she 'should deeply regret that the Observatory should think that I had anything to do with starting this controversy.' She had tried to make clear with the journalists who had come to call that she was in no position to make statements about the work of the Royal Observatory. The man from the *Maidenhead Advertiser* said he 'found Miss Belleville to be a well-informed lady, communicative, and quite an expert in calculating time . . . The only approach to reserve was when questioned relating to the Observatory and matters closely connected therewith, remarking that without the cognisance and sanction of the Astronomer Royal it would be unwise to make any information public.' She knew the importance of keeping Christie on her side, remarking that 'I cannot speak too highly of the kindness shown me by the Astronomer Royal.'

Ruth Belville had been far too young to know first-hand the trouble her mother had with George Airy in 1856 after she was widowed and the astronomer had accused her of breaking and entering. She was only two years old, but the story will doubtless have been passed on many times over the subsequent years, always with the same moral attached: he lets us in as a favour; we've no official connection. Don't do anything that might annoy him! Christie had taken up the post of Astronomer Royal in 1881 and was by 1908 just two years from retirement. While he was not the autocrat that his predecessor was, nevertheless the women were only too aware that their livelihood rested on the kindness of the men at Greenwich. Rocking the boat was not a good plan, and unexpected let alone factually incorrect newspaper articles being splashed around the London dailies was probably not the best way to keep things on an even keel, as the *Daily Graphic* coverage of 1892 had shown. As she travelled to Greenwich that Monday morning, Ruth must have been in turmoil. What she needed, perhaps, was a tonic.

St John Winne and his cronies seem to have helped with that, too. If the Standard Time Company was trying to get the idea of synchronized clocks firmly into the minds of the general public by their 1908 propaganda campaign, they succeeded wholeheartedly. Readers of the *Manchester Guardian* on 8 May 1908 – two months after Winne's lecture – were faced with an extraordinary advertisement for Beecham's Pills. 'You will have noticed that a clock left to itself is rarely right; it requires to be regulated carefully. If it is two minutes slow now it will be a quarter of an hour behind next week, and it is never two days alike.' The reader was by now feeling pretty unhealthy, as the eye was drawn to the big capital letters just a few lines up: 'SLUGGISH LIVER, SICK HEADACHE, LOSS OF APPETITE, INDIGESTION, CONSTIPATION.' Go on then, thinks the reader, tell me why Beecham's Pills and unregulated clocks go together. 'We ourselves are very like clocks. Only in rare instances can we be left to go by ourselves. To keep Greenwich time – always to be right and never to be wrong – we want to be regulated periodically.' Come now, scoffs the reader, that's a pretty tenuous analogy. Tell me again how constipation is like Greenwich time? 'A little neglect and the stomach will get hopelessly slow . . . it will be much worse than an irregular clock: this only causes you to lose your train; an uncertain stomach will make you lose your work.' I see, the reader might have mused. I need to keep my *clocks* regular so I don't miss my train – Mr Winne has been telling me that all year – and in *exactly* the same way, I need to keep my, ahem, *movements* regular otherwise I'll *lose my job*! This is terrible news. Hand me the Beecham's Pills. STC's propaganda campaign must have prompted this bizarre bit of advertising, which indicates the extraordinary renewal of public attention Winne had brought to bear on Greenwich time with his scheme.

But was Ruth right to worry? Actually, we don't know whether she even saw let alone spoke to the Astronomer Royal that morning. She probably didn't. The watch was checked as usual, and Ruth received her certificate stating how many seconds and tenths-of-seconds it differed from the time by the master clock. The photographer hung around patiently until Ruth emerged, and then gathered the party into position in front of the Shepherd clock dial to get his shot for the *Daily Express*. As he went back to Fleet Street, Ruth went on her rounds, and there's no suggestion any of her customers cancelled their subscriptions. In fact, it's possible she gained a handful more. Further press articles appeared, the world kept turning, the clocks kept ticking and it seems that the Astronomer Royal didn't mind. He later said that he 'wished every facility to be afforded her', which prompted Ruth to proffer 'apologies for all the trouble I give you and wish many thanks for the kindness of the Observatory staff'. Despite Winne's efforts, the Belville business had weathered another brief storm.

We know all about STC's customers, because STC was in the business of promotion. Ruth Belville was altogether more private, more discreet. She was the Savile Row to STC's Oxford Street, the Pall Mall to their Leicester Square. She never said much about her clients and they appear to have kept pretty quiet about her too. Nevertheless, the clues are there. We don't know much for certain, but with the newspaper reports of the day, her own correspondence, a little detective work and some educated guesses, plus lots of maps and exploration of what's on the ground a century later, we can piece together a picture. In 1908, Ruth still had enough customers for a viable business. She wanted more, but since her father's day the numbers had been dropping off significantly. This is no surprise. As we have already seen, in John's day there was no easy way to get access to Greenwich time, whereas Maria and Ruth both had to compete with the electric time signals from the Observatory and from STC's Queen Victoria Street control station. The new technology had hit the Belvilles hard, as Ruth indicated in a letter to the Observatory a couple of years later, once the dust had settled. 'I wish I had the number of firms my father had – over 200, I think. I can well remember my mother visiting over 100 firms – nearly all in the City, Minories, Clerkenwell and London Docks – but most of the firms were close together.'

The synchronizing technologies of STC had clearly affected the Belvilles' bottom line, and there is quite reasonable bitterness in Maria and Ruth's writings. To one journalist in 1908, Ruth said that her chronometer 'is the most reliable method of providing the accurate time, and realising this many of the best firms will have nothing to do with synchronised time'. Then in about 1910 she observed, 'I think it was in the early seventies that Time current commenced to be given by Electricity . . . This was naturally a great vexation and a loss to my mother as some of the firms began to be synchronised . . . I am naturally prejudiced against synchronisation.' By 1913, she had become even sharper in her criticism of the technology. Writing in *The Observer*, she noted, 'As to syn-chronised clocks, doubtless they are of service to the general pub-lic, and possibly to those who sell cheap watches . . . but to the high-class scientific watch and chronometer-maker, Greenwich mean time is required to tenths of seconds,' an accuracy only achievable via her chronometer service. In the thick of the 1908 press frenzy, though, she could still be bullish about her business. She told one reporter who visited her on that fateful Friday evening that her business was 'still comparatively flourishing. Many of the foremost chronometer manufacturers in the City are still glad to have my meantime.' Her faithful customers were pret-ty numerous, she said, and 'the round extends from the City to Kensington, and from Baker-street to Chelsea.' In the *Kentish Mercury*, she also mentioned customers additionally in the Borough (south London) and Hatton Garden, where her mother lived when she married John Belville. This was quite an extensive area to cover each Monday.

The routes she followed were never properly written down. It can only be a matter of conjecture exactly where she travelled every week, but we can work out a rough idea. A newspaper article written in 1929 described a typical day with Miss Belville. 'From Greenwich Observatory Miss Belville does her round of firms, call-ing at large industrial undertakings which have found it essential to have the correct time without fear of contradiction.' Let's assume these are in the London Docks she already mentioned, on the north bank of the river near Shadwell, and the thriving heart of London's trade. 'Next the watchmakers in the Strand and the City are visited, and at the conclusion of a long tour Bond-street, Regent-street, and Mayfair firms are called on, so that fashionable shoppers may set their watches by Greenwich . . . Altogether about fifty clients, including the houses of two millionaires, are visited each week.' Ruth Belville was, by profession, a creature of habit. The noted horological writer Donald de Carle interviewed her a few years before she died, and we are lucky that he published a report of their conversation in his book, *British Time.* 'She usually called on the same day each week and at about the same time and carried, in addition to her handbag, a shopping bag, no doubt collecting her shopping during her journeys.' After all, she called at the fashionable shops in London's West End, so she might as well take advantage.

Yet in all of the reports that have been found so far, only one mentions any of her customers by name. It seems that Ruth extended to her local newspaper, the *Maidenhead Advertiser,* a rather more privileged interview than she granted the London dailies. It's not much, but it gives us a valuable glimpse into the daily life of the Greenwich time lady. 'My clients', she told their reporter in 1908, 'include John Carter; Moore, Atkins, and Brockbank; Parkinson and Frodsham; J. W. Benson; John Walker & Co.; and Mappin and Webb.' As mentioned earlier, Maria Belville was given a pocket watch in 1882 by Parkinson and Frodsham, though neither she nor Ruth ever mentioned it. Perhaps this was a bit of product placement by the renowned chronometer-making firm. The John Walker company, a prominent supplier of clocks for the railway industry, we have also met in connection with the publicity shot of Maria Belville on her retirement in 1892. By 1908 it was trading from Fenchurch Street and New Bond Street. Brockbank and Atkins we met earlier, too. They had that 'different story' about the true origins of the chronometer time service. All these firms were in and around the City of London, except Mappin & Webb, which had stores in Oxford Street and Regent Street, and Benson, which besides their City office had an Old Bond Street store trading as 'Jiggumbob'.

Mappin & Webb we will encounter later but this 1908 article clears up a problem. Forty firms is a lot to visit in one day, even for someone 'hale and hearty', as Ruth was described in one of the reports. It had always been assumed that she must have done her rounds over at least two days each week, although if that was the case, her Maidenhead years must have been a real trial given the length of her overall journey. But now we know what really happened. 'How many firms have you to whom you take the time in London?' asked the *Maidenhead Advertiser*'s reporter. Ruth replied, 'About forty. But I do not visit these every week; some I call upon once a fortnight. On an average I make about thirty calls each Monday after visiting Greenwich, and it is a hard day's work.' Hard indeed, especially with that shopping bag, but of course, unlike her parents, she now had extensive public transport to help. While her round certainly involved a lot of walking, she was, according to a 1913 newspaper article, 'carried all over London in tram, bus and electric train'.

Given all this evidence about the daily routine of the Greenwich time lady, we're now in a position to get the period maps and guides out and work out how she fitted it all in – and how it looks today. First stop, then, to the London Docks. She had two choices. If she had clients in Rotherhithe or the Surrey Docks, she could catch a tram from Greenwich to New Cross before taking the East London Railway north to Shadwell for the London Docks. Or, if she had customers in the Millwall or West India Docks, she would need to get onto the Isle of Dogs. Let's explore the second option. A fifteen-minute walk down the hill from the Observatory, through Greenwich Park, takes you to the river Thames at Greenwich Pier. At the turn of the century, the only transport link across the river to the Isle of Dogs had been the steam ferry that plied between Greenwich Pier and North Greenwich railway station on the north bank (above today's Island Gardens station, and emphatically not where today's North Greenwich tube station lies, now on the south side of the river). The ferries were reasonable enough in good weather but in winter fogs the service was unreliable and often dangerous, so in 1902 the London County Council dug a pedestrian and cyclist foot tunnel under the Thames, linking Greenwich with the Isle of Dogs, enabling the labouring population of run-down Greenwich to reach their work in the docks. The characteristic glass-domed brick entrance buildings still stand proud, because the tunnel, together with its slightly younger sister down-river at Woolwich, is still very much in use today.

Having walked through the tunnel, Ruth would have arrived at the Isle of Dogs, then home to the sprawling complex of the Millwall, West India and East India Docks. Today, these have largely been filled in and the docklands area is home to London's new financial district. In Ruth's day, the quickest and most convenient way to reach the London Docks was the Millwall Extension Railway from North Greenwich, at the southern tip of the Isle. Trains ran every fifteen minutes northbound to Millwall Docks and South Dock stations, before turning sharply west onto the London & Blackwall Railway, passing West India Dock, Limehouse and Stepney stations before reaching Shadwell. Here she could alight for the

London Docks and her first customers. These were the streets that were close to some of the Whitechapel killings in the 1880s, and would have borne bad memories. On returning to the railway, Ruth could continue a couple of stops west to the terminus at Fenchurch Street, near the Tower of London. Sharing the carriages with Ruth would have been a rich assortment of dockyard workers, sailors, craftsmen and the streams of messengers carrying orders between City shipping offices and the ships in the docks, although this trade was drying up with the development of the telephone network.

The route followed by passengers on these lines in 1908 is almost exactly the same as that of the modern-day Docklands Light Railway (DLR), opened in 1987, which reuses many of the old railway alignments and viaducts. The first DLR London terminus was at Tower Gateway, a stone's throw from Fenchurch Street, and these modern driverless computer-controlled trains seem a world away from the steam-hauled service in the age of the docks, but the new technology echoes the old more than we might think. In fact, the London & Blackwall Railway (L&BR), opened in 1840, saw the first commercial use of the electric telegraph anywhere in the world and its early operation relied entirely on electric signals passing alongside the tracks. Steam locomotives were a huge fire risk on this line, which ran beside docks bristling with timber sailing ships and warehouses crammed with flammable stock, so the L&BR was cable-hauled for its first nine years. Yet it managed to operate a four-trains-per-hour service in both directions. Contemporary illustrations show clearly the electric telegraph instruments used at each station in a complex and sophisticated signalling system to control the operation of the stationary steam engines at each end of the line. These drove the seven-mile-long cables that the trains were clamped onto. Now, the DLR glides along the old brick viaducts of the L&BR, built almost 170 years ago, but run by sophisticated systems and powered not by a moving cable, but by a live electric rail. The physical pathways have stayed the same but the context has totally changed. Industry and merchant trade have given way a century later to high finance.

Yet echoes of the past keep sounding. Let's assume Ruth Belville has transacted business at her dockside customers and has reached Fenchurch Street. Next, she tells us, is the Minories, the City and Clerkenwell. The Minories comes first, a thoroughfare beside Fenchurch Street station running north from the Tower of London. Her clients might have been nautical instrument-makers – of which it long held a number – or Jewish brokers and dealers who worked there and in neighbouring Houndsditch. It was here, in Mitre Square, that Catherine Eddowes's body had been found in 1888 and Ruth's third cousin, Thomas Lewes Sayer, had viewed the corpse in the City Mortuary in Golden Lane: a bit close for comfort, time to move on. A brisk walk, or a ride on the underground railway from Aldgate up to Farringdon, would bring Ruth into the Clerkenwell district, a long-standing one of clockmakers, watchmakers and jewellers. Once done, she could then return via the underground to Moorgate and walk south into the City, the home of commerce and money-making with its innumerable banks, exchanges, brokers, merchants and firms of every description. The General Post Office was headquartered there, as was the Bank of England, and it encompassed London's press and its legal establishment. Plenty of demand for the right time. Many of her customers were clustered in the streets of the City.

She was now close to Fleet Street and its westward continuation, the Strand. It is easily walkable, but if the weather was foul or Ruth was tired, there were plenty of omnibus services (mostly horse-drawn, although motor buses were starting to appear) throughout central London. For example, the white-coloured 'Walham Green' service, route five, operated every six or seven minutes between Liverpool Street and Fulham in the south-west, taking in the City and Strand along the way. Short-hop journeys would cost only a penny or two and services ran from eight in the morning right through to midnight. Alternatively, a network of electric tramways was spreading around London in the Edwardian decade: fast, cheap and convenient.

Before the Strand, though, Ruth needed to pop over London Bridge to the Borough, to see a customer there. It's a bit of a walk, so she would likely have jumped on a route 90 bus (dark green) from London Bridge south down Great Dover Street, hopped off, transacted her business, then climbed onto a bus in the other direction back to the City. They came about every five minutes, so this detour wouldn't take too long.

Then she could go to the Strand, after which the logical next stop would be her customers in Kensington and Chelsea, paired together in south-west London and reached easily by the underground from Embankment or Temple stations to South Kensington. Nearby Chelsea could be reached on foot, and on returning to South Kensington, Ruth could use the second of the brand-new electric tube railways – the Piccadilly, opened in 1906 – to reach a station called Dover Street (now Green Park), which is right at the bottom of Bond Street, with Regent Street running in parallel. Ruth could visit her fashionable customers including

Mappin & Webb and 'Jiggumbob', collect any West End shopping (meaning she did not have to lug it about town all day) before heading to Oxford Circus to pick up the Bakerloo tube north to Baker Street, her final customer stop. Ruth would walk back to Paddington (it's hardly worth taking the tube for just one stop) and work for the day was done, many hours after leaving the Royal Observatory. This was not the end of it, of course. She had to get back home to Maidenhead, a further hour, plus waiting time, before she was sitting with a cup of tea in her St Luke's Road cottage.

All this made a long enough day, but by 1910, Ruth had moved to the tiny lodge cottage at Ockwells Manor, on the out-skirts of Maidenhead, and the journey from there to Maidenhead railway station can easily take forty minutes to walk. Ruth Belville may have been hale and hearty, but this house move was apparent madness, given the location of her employment. Her journey from home to the Observatory every Monday could now have taken three hours, including waiting around for all the connec-tions, and one hour of that was spent striding along on foot. And this didn't include all the travelling and walking for her subscriber round! When she said it was a 'hard day's work', she certainly meant it, so what did Ockwells Manor have that St Luke's Road didn't?

It's a delight, it must be said, and the helpful *Maidenhead Advertiser* ran an article about it in 1908. A manor was established on the site, to the south-west of Maidenhead, in the thirteenth century, with the present house built in the reign of Henry VI. It is renowned for its stained glass and is surrounded by rolling fields. Its tiny little white-painted lodge became Ruth Belville's home for a couple of years and must have been a quiet retreat from the urban stresses and strains of her Monday job. It still might be; the lane outside remains quiet and rural, frequented by dogs and walkers, and the fields are a chocolate-box picture of bucolic picturesque. But the walk to Maidenhead has changed more than a little since Ruth's weekly promenades. If you stand listening outside Ockwells Manor today you can hear the usual country sounds: birds singing, cheery golden retrievers barking at the passing squirrels – and a distant low rumble, the ever-present sound of vehicles roaring along three nearby motorways which come together at a junction just across those fields. In order to reach Maidenhead from Ockwells today, Ruth's walking route becomes a footbridge over the A404 carving its path towards High Wycombe and the M40 to Oxford.

Ruth didn't stay at Ockwells for very long. In 1911 or 1912 she upped and moved inwards to Ewell, in Surrey, and it was not just Ruth who was being uprooted. The world was soon to change, drastically and forever. Local and global were in constant collision in the decades before and after the Great War. Before we leave Maidenhead, though, let's hear once more from Ruth Belville as she discussed time with the local newspaper reporter, that difficult day in 1908. 'Of course you find that the Maidenhead clocks keep excellent time?' The reporter was about to leave and threw in this final question. Ruth's response was sharp. 'Hardly. Some of them keep good time, but others do not. Please do ask the Town Council to see to their clock on the Town Hall. It is often wrong, and I feel very cross about it when I am on my way to the station.' That town hall is long since demolished, but there is a very nice four-sided brick clock tower just outside the railway station forecourt. How well does it keep time today, one century later? Spot on. Perhaps the town council heard Ruth's request after all.

Chapter 6

SUMMER TIME AND RELATIVITY

{ 1908–1920 }

William Willett and his daughter Gertrude at home, about 1895.

NEW BRIDGE STREET, CITY OF LONDON, 21 MAY 1916. New Bridge Street, near St Paul's Cathedral, leads off Fleet Street, home of London's newspaper trade. The reporters were out in force on 21 May 1916, sniffing out stories and anecdotes, anything to capture the mood of the nation that day, because at two o'clock in the morning, Britain's time had changed. One reporter took the right turn into New Bridge Street and came across some pavement graffiti, chalked onto the flagstones of this busy thoroughfare connecting Blackfriars railway station with the bustling City of London. 'All Fools' Day, May 21. Get up one hour earlier and kid yourself you haven't.' Who were the fools? The *Times* journalist smiled, scribbled down a few notes, and continued his work piecing together a long article for the following day's issue, under the headline, 'The Extra Hour: First Summer Time Day'.

That morning, 21 May 1916, two o'clock had become three o'clock and Britain had entered its first day of Daylight Saving Time

(DST). People have become so used to setting their clocks one hour forward in spring, and returning them to standard time in the autumn, that we can find it hard to imagine a world without DST. It began in 1916 but that event was as much the end of something as the beginning. It was in the Edwardian decade, the first ten years of the twentieth century, that time changed for Greenwich time, and for much else. We now know that Ruth Belville's world was rocked in 1908, by the Standard Time Company's attempts to win market share, and we've seen the wider context of clock synchronization that underpinned that 'lying clocks' campaign. But that was just the tip of the iceberg. What was really going on?

As mentioned earlier, it was the American writer, scientist and politician, Benjamin Franklin, who coined the phrase 'time is money' in his 1748 tract, *Advice to a Young Tradesman*. Franklin was advocating industry and frugality as the twin means to richness. The phrase was certainly remembered for a long time, coming into its own at the turn of the twentieth century when some of his writings were republished. The young tradesmen were not the only ones to benefit from Franklin's temporal wisdom; his homily struck a chord with Edwardian industrialists and businessmen too. It is often said, today, that the idea of DST – of changing the time on our clocks and watches during summer months in order to get up earlier – was first conceived by Franklin, in an anonymous letter to the *Journal de Paris* published on 26 April 1784. Writing at the age of 78, Franklin was reiterating his beliefs about the value of thrift, as his letter discussed the Parisians' love of rising late and their concomitant reliance on artificial lighting late in the evening. The solution was to rise earlier – in fact, to rise when the Sun did, and therefore at varying times through the year. 'Every morning, as soon as the Sun rises, let all the bells in every church be set ringing; and if that is not sufficient, let cannon be fired in every street, to wake the sluggards effectually,' he thundered.

Franklin's tongue was firmly in his cheek as he wrote his letter, which was a parody on the scientific method and on his own stern writings on the morals of economy, but his underlying message of thrift rooted in soberly industrious working patterns was deadly serious. To be thrifty, to be morally strong, citizens should make use of natural rather than artificial light whenever the seasons gave them the option. Franklin did not conceive DST in this letter. He simply suggested people get up earlier and go to bed earlier in summer, when there is more daylight to be had. The idea of a technological method by which this behaviour could be produced – the government-sponsored changing of time every summer, now called DST – was to come much later. The famous clock-shifting solution came in 1907 from a London house-builder, William Willett.

William Willett was born on 10 August 1856, exactly four weeks after John Belville passed away at home in Greenwich. Willett joined his father's building firm when he was in his early twenties and they spent the subsequent quarter-century building distinctive houses all over London and Brighton. His father retired in 1903 and William began to diversify the building firm's reach into areas including estate agency, property speculation, building technologies and mining. But true fame came to William Willett with his proposals for changing time during summer months: DST, or British Summer Time as it is known in the UK. Many years after William Willett had died, his daughter, Gertrude, described the circumstances in which her father had come up with the idea. 'Every morning before breakfast he rode through Petts Wood . . . There were beautiful bridle-paths leading through the woods under pine trees; it was here that he first thought of Daylight Saving.' Gertrude would have known better than anyone alive: she used to accompany her father on those horse rides, starting from the family home in Chislehurst (which her father's firm built). The portrait at the start of this chapter was taken by Edgar Davey Lavender, a Bromley photographer specializing in equestrian subjects, and it shows the pair of them on horseback ready for a morning ride. Willett's concept of DST was simple. He had been incensed at the 'waste' of useful daylight on early summer mornings. Though the Sun had been up for hours, people were still asleep. How do you persuade them to get up earlier in the summer to make better use of daylight? Willett proposed that all clocks be put forward, so people would think that they were rising later than they actually were. This was a more palatable step forward from Franklin's exhortation that people should get themselves up at sunrise.

In 1907, following a maiden speech to the Metropolitan Public Gardens Association in London, Willett published a pamphlet called *The Waste of Daylight* outlining his plans. Under his original scheme, Britain's time would be advanced by twenty minutes on each of four successive Sundays in the spring (totalling eighty minutes), returning by a similar set of four twenty-minute steps in the autumn. His first positive outcome was to get a daylight-saving bill sponsored in Parliament by Robert Pearce MP, who was an immediate convert to his ideas. The bill was referred to a Select Committee to examine its contents. The committee first met on 5 May 1908 and for the following two months, every aspect of Edwardian timekeeping was placed under the scrutiny of the nine men nominated by Parliament to consider the clock-shifting scheme. Of the forty witnesses called to give evidence besides Willett himself, three men testified who at the same time were changing Ruth Belville's world. One she knew personally: Sir William Christie, the

Astronomer Royal at the Royal Observatory, the man who continued to give her access to Greenwich time every Monday morning. A second man she knew not at all, but would encounter fifteen years later – Frank Hope-Jones, owner of an electric clock firm with interests in wireless communication. But the third, who turned up to give his witness statement on Tuesday 26 May 1908, was known to Ruth Belville only too well: St John Winne was not finished with the lying clocks.

Here he was, the colourful company promoter, taking his seat in the committee room before five prominent Members of Parliament that Tuesday in May, only a few weeks after his lecture to the United Wards Club, the effects of which Ruth Belville was still coming to terms with. Winne, the man who wanted above all to secure legislative support for a time system based on electrical synchronization, standing in front of a committee which could change Britain's timekeeping culture forever. Winne had stepped into the big league: this was the speech that could turn around the Standard Time Company for good. Sir Edward Sassoon was in the chair. Member of Parliament for Hythe, on the south-east Kent coast, Sassoon was a prominent businessman who had married into the well-known Rothschild banking dynasty. And he knew a little about the modern politics of timekeeping, having a keen interest in state control of cable telegraphs overseas for the service of Britain's Empire, with their implications for global time systems.

First to the stand that day had been William Willett, who had returned to tell the committee about some new letters of support he had received for his scheme since he was last interviewed. He read them out and withdrew from the room. Next came a quick interview with Dr Roberson Day, a physician, who described the health benefits that he felt would accrue to the population if they experienced more natural (as opposed to artificial) light in summer. Sunlight would bring benefits incalculable to Britain's people and future, he seemed to be saying, 'and if we could rise early in the morning and have more of the sunlight, there would be less anaemia, less of the rickets, and it would tend largely, I maintain, to prevent the deterioration of the race, which is now becoming such a serious problem.' Timekeeping was not just a matter of business and economics. It was a matter of health, of hygiene, of racial vigour, and this made it the interest of the state. After the aptly named Day had withdrawn, it was time for a retired admiral, Sir Edmund Fremantle, to spend considerable time confusing everybody – especially himself – as to his opinions on the fine detail of Willett's scheme. As a man familiar with the role of Greenwich time for navigation, Fremantle could have offered useful insights into the situation as it might have affected the chronometer-makers of the City and the dock area customers in Shadwell receiving their weekly visit from the Greenwich time lady. As it turned out, though, Fremantle merely tied himself into knots with complex calculations and insults about the inhabitants of the West of England (it was all about local time-differences but he made it personal).

With this slightly baffling exchange out of the way, the chairman called in St John Winne, who had been waiting patiently. He got straight to work. 'The Standard Time Company has no monopoly whatever, but happens to be the only company in London at the present time that has wires radiating over a great portion of it for the purpose of correcting clocks hourly and uniformly with Greenwich mean time, and that we call "synchronising".' The point being explored by the committee with Winne was the practicalities of changing the clocks twice each year – or, if they went with Willett's proposal, four times forward every spring, and four times backwards in the autumn.

Was there a technological solution to the problem of knowing the right time on all these mornings? Not at all, explained Winne. Every individual simply needed access to sufficient public clocks that were set to the right time; they could then set their own clocks and watches accordingly. Does this sound familiar? 'What we want is a sufficient number of reliable public standards . . . not, as they do to-day, showing all sorts of times, so much so that they get called "lying clocks" and create an outcry with regard to them.' In order for all publicly visible clocks to be reliable, Winne observed, 'means should be taken to ensure that they show correct time, and if synchronisation is the only way of making them show the correct time, they should be synchronised.' With Willett's proposals of shifting time eight times each year, Winne saw a golden opportunity to press home his demands for legislation on public clock synchronization. Public timekeeping had taken on a new dimension, and his company's technology was ideally placed to take control of the clock.

This really was the opportunity of a lifetime. Winne had an appreciative and highly influential audience of men who could turn round his company's fortunes at a stroke, and he milked the opportunity for all it was worth. 'I have an ordinary synchronised clock here; it will only take two minutes to put it on the table, if you like to see it work.' Yes, they would like to see it. He set it up; he was now able to describe in minute detail the technical means by which his company could effect the twenty-minute time-shifts automatically from their central station in Queen Victoria Street. It was tricky, because their technology was designed to make small corrections, not large changes of time. But he had worked out a compromise. It could be done.

Thomas Richards, MP for West Monmouthshire and General Secretary of the South Wales Miners' Federation, cut in. 'We had a witness here a short time ago who like yourself is a very practical man, and he made a statement which is rather contradictory to your evidence.' It had been 60-year-old Thomas Wright, a clockmaker and teacher born and raised in the horological heartland of Clerkenwell. He had been representing the traditional clockmakers of the British Horological Institute – the clockmakers who had been so put out by Winne's 'lying clocks' exchange, the old trade fiercely opposed to electrical synchronization. Wright's opinion of Willett's idea was that the principle was good – getting up earlier to make better use of daylight – but the four-times-twenty-minutes changeover was not. 'I think it is a very stupid suggestion,' he had said to the committee. A single one-hour change was better. 'I strongly object to changes of the clock except by complete hours,' he had told them. Willett's idea, while laudable in principle, was in practice another attack on the mechanical horologist's trade, thought Wright. Like Winne's electrical synchronizers, Willett's seasonal time-shifts were knocking about the hands on carefully crafted timekeeping machinery. 'Every watch in the Kingdom would have to be altered. You cannot do that by electricity; it would have to be altered four weeks in succession. People would have to say: "Let me see, is this the week that I have to alter my watch; how much does it differ?" It would be shocking. It would be no good a man having a good watch.' It would spoil a good watch, would it? 'A good watch would be of no use if you had to alter it eight times a year,' he agreed. 'A man who buys himself a watch which keeps time to a minute a year would have to shift it about eight times a year. It would condemn it . . . I am quite sure the majority of people would find it an intolerable nuisance.'

Wright admitted that he was wearing a watchmaker's hat as he spoke these words, and Richards went on to suggest that Willett's proposals might actually be good for the watch trade. 'We do not want any help of that sort,' Wright had retorted. Put the clocks forward one hour, once, and leave them there. Move to mid-European time, that was his proposal. Here, things had gone awry for watchmaker Thomas Wright, as the committee had rather acidly interrogated him as to which countries in the world would be left keeping time by the Greenwich meridian, and he couldn't remember. 'It goes through the northern part of Africa,' he had offered, confused. The committee asked, 'What would North Africa be doing about it?' Wright had muttered, 'I do not know whether they would pay any attention to it,' and withdrew. In the end, history shows that Wright's view prevailed, and we now change our clocks in whole hour steps. Indeed, some claimed this development for Wright alone, his 1933 obituary stating, 'We are indebted to Mr Wright alone for the conception and later acceptance of the complete change of one hour in the Willett Daylight Saving Scheme, which would not have worked in any other way, and millions must thank him for that benefit.'

Wright's appearance before the daylight committee had been the previous week. Committee member Thomas Richards continued his examination of St John Winne, explaining Thomas Wright's views on changing the clocks. 'The question put to him [Wright] was: "You do not see any practical difficulty?" And his reply was "No; one suggestion is to make the change at two o'clock on one of the Sundays. I do not know how you are going to get over that; who is going to alter the station clocks."' Richards continued to recall the conversation: 'It has been suggested that it might be done electrically – electrical synchronisation.' Wright had been prompted; his answer had been, 'Yes, that might come in time, but at the present time not one clock in a hundred is so arranged. It means a big expense, and it is unnecessary, because a good clock does not want synchronising.' Winne was itching to respond to the clockmaker's arguments – after all, he had spent months trying to demolish the trade as old-fashioned and reactionary. This was his chance to press his points home with the parliamentary committee, and to do so he was about to attempt to hijack Willett's bill: 'I recently lectured on this subject on "The Time of a Great City," and my lecture was a plea for uniformity of public clocks . . . if I could see any change effected, I should like to see this Bill running for both objects at the same time.' It made perfect sense. In one piece of legislation, timekeeping could be transformed from laissez-faire individualism to state-controlled uniformity – with STC, naturally, providing the centralized controlling infrastructure.

The old-guard horology trade was standing in the way of all this. Winne started to twist the knife. 'From time to time there have been public outcries against "lying clocks" . . . on every occasion when that outcry is made the clockmaker always says: "Synchronising! Synchronising is an insult to a good clock; it does not require it"; but while he says that about synchronising he never suggests a remedy, and it goes on the same as before, year after year . . . It is always the cry of the British horologist; he is always against any progress.' It was humbug, he said, that altering clocks harmed them; 'instead of synchronising being an insult to a good clock it gives just the touch of perfection which it lacks.' Concluding with a diatribe about the urgent need for a timekeeping authority in Britain, charged with synchronizing public clocks around the country, Winne repeated his proposal that Willett's bill should go much further than it did and left hanging in the air the thought that the synchronizing authority might be a private company – like his. He too then withdrew.

Winne's promotion of STC's electric time business may have been shameless but, in his support for Willett's bill, and in his advocacy for a technical solution to the problem of clock-setting, he was not alone. The witness called to speak after Winne left, electric clock impresario Frank Hope-Jones, backed Winne's call for the use of electricity and his criticism of the horology trade, saying, 'if I might reinforce what I heard Mr Winne say just now, that the difficulty raised by the trade, the somewhat old-fashioned trade of mechanical clockmakers – is (not to put too fine a point upon it) a "bogie".' The knives were out for the old guard still keen to keep clock-setting as a human intervention, not an electro-mechanical one.

The other witnesses were called, one by one. Some were in favour of Willett's daylight-saving plan, some were against; most had mixed feelings. Over ten separate sessions the committee discussed the proposals with the invited speakers, and at the penultimate hearing, on 18 June 1908, Astronomer Royal Sir William Christie was called. He was the guardian of Greenwich mean time, the keeper of the clocks, so what did he make of Willett's time? Not a lot. 'My view is that this ideal of early rising may be secured in other ways.' Time was time; if people wanted to enjoy the summer daylight, they could get up earlier. Their employers could change the working hours. After all, he had; at the Observatory his staff both started and finished work earlier in summer than most workers in clerical and commercial firms, so they had the benefit of lighter evenings to play for the Observatory's hockey team. Change the hours by all means, but why change the clocks?

In the end, the 1908 committee went for the compromise amendment. They were convinced that Willett's ideas were sound but they disagreed with the four-times-twenty-minutes detail of his proposals. The report they submitted to Parliament put forward the plan to advance the clocks by one hour in April, returning to normal in September. It was a compromise, but it wasn't the permanent advancement to mid-European time advocated by Thomas Wright and others. But after all this, the Prime Minister, Herbert Asquith, was against the plan and, without his backing, Parliament rejected the bill. Not to be deterred, Willett gained support for a further bill the following year, this time sponsored by Thomas Dobson MP. It too was referred to a Select Committee, which met sixteen times between March and August 1909 under the chairmanship of James Tomkinson MP. Winne was afforded no opportunity of speaking before this committee, and neither was Hope-Jones, Christie nor any of the other 1908 witnesses with the exception of Henry Babington Smith, Secretary to the Post Office, whose previous evidence had been judged so useful that he was given a second audience. This committee was much less positive about Willett's ideas, and at the end of their deliberations, their recommendation to Parliament was that 'the Bill be not further proceeded with', notwithstanding their approval of the principles behind Willett's proposals. For them, the shifting of summer working hours earlier in the day was a laudable aim that should be achieved by voluntary means, not by changing the clocks.

Willett was indefatigable. Despite getting support for two daylight-saving bills, by 1909 his ideas were no further towards the statute book than when he first conceived the idea while riding through Petts Wood. In 1911 he brought another bill, and in the following three years a further two bills. All were defeated. It was now 1914. The world was at war, and Willett decided to call it a day on his summer-time scheme. In 1915, he travelled to Spain to visit the Matallana Company, a bauxite mining firm in which he had shares. He had heard of fraudulent activity in the company and wanted to sort it out. On his way back, he developed influenza. He made it home to Chislehurst and on 4 March 1915 died there, aged just 58. He was buried in the

Chislehurst parish church of St Nicholas. His grave has recently been restored, following renewed interest in the man who gave us 'summer time', and it happens to be just a few footsteps from the final resting place of John Arnold, the chronometer-maker who constructed Ruth Belville's beloved timepiece back in 1794.

The irony is that the Great War was the very catalyst Willett's ideas needed to be made into law. He had been pushing his scheme not just in Britain but around the world: in continental Europe, the Americas, Australasia – anywhere a reasonable distance from the equator which would benefit from shifting working hours during long summer days (at and near the equator the problem does not arise). And it was this international advocacy which, ironically, was to see his ideas posthumously taken up in his home country. The very last letter Willett received following his tireless championing of DST abroad was from a correspondent in Germany, in June 1914, who promised to lay his plans before the German Emperor – who knew the correspondent 'personally and very well'. That last letter, written just before the outbreak of war, was the one that may finally have prompted change. A year after Willett's death, British politicians were horrified to learn that Germany – the enemy – was indeed trying out Willett's time-shifting scheme. Germany's *Sommerzeit* (summer time) began in April 1916 and was swiftly followed by schemes in Austria-Hungary, Holland, Belgium, Denmark and Sweden. British commentators outlined wartime benefits, such as saving fuel and prolonging daylight working in the shipyards: DST would enable the yards to stay open later in the evening before the artificial lighting was switched on. After years of Willett banging his head against the Parliamentary wall, the one after his death saw the Summer Time Act passed, and 21 May 1916 became the first day of Britain's Summer Time.

Willett had won. He had changed the times. His original plans, based on moral outrage, were deemed unacceptable in peacetime, but in war they took on a new guise: as fuel-saver and as a weapon against the enemy, although ironically it was the enemy that adopted the idea and made the first move. Willett had shown that timekeeping – as shown on our clocks and watches – could be fluid, that it could be changed depending on the point of view. He had demonstrated that Greenwich time was not sacred but a human construction, as artificial as his own daylight-saving time. He had offered his own interpretation of a concept which was taking the scientific world by storm: that time was relative, that it all depends on where you are standing.

Time was relative. Change was in the air and the change was led by Germany. While Germans were setting their clocks forward in April 1916, they were also trying to come to terms with an even more radical shift in their conceptions of time. It would be events after the war's end, in 1919, that would propel the scientific theories of relativity into the public consciousness. By the early 1920s every popular medium, from newspaper to film to poetry, would be grappling with the consequences of the new scientific order. But it was in 1915 that Albert Einstein, working in Berlin, presented his full theories on relativity to the German scientific community; and it had been further back, in 1905, as Willett was riding his horse in Petts Wood, that Einstein had published his first revolutionary paper setting out his basic ideas on relativity and revealing the effects his theory would have on the nature of time.

The conceptual connections between Willett's DST and Einstein's relativity were not lost on the public. Relativity hit the headlines in 1919 after astronomical experiments carried out on the islands of

Principe (off western Africa) and Sobral (in Brazil) examined an eclipse of the Sun, demonstrating, the scientists claimed, that Einstein's theories were correct. Light, they suggested, had weight, just as matter has, and therefore it could be bent by the gravity of large bodies like stars or planets. The *Daily Express* reported the news on 8 November 1919, with an article penned by Charles Davidson of the Royal Observatory, who was in charge of the Sobral expedition. Its headline read 'Upsetting the Universe: Dizzy Results of the New Light Discovery', and in the piece, Davidson explained that, 'The result . . . stated in non-technical language, is to prove that light has weight in proportion to its mass, as matter has; but it takes an appalling amount of light to weigh an ounce. The cost of light as supplied by gas and electric light companies works out at something like £10,000,000 an ounce. This points the moral of daylight saving. The Sun showers down on us 160 tons of this valuable stuff every day, and yet we often neglect this free gift and prefer to pay £10,000,000 an ounce for a much inferior quality.' Einstein was upsetting the universe, Willett upsetting the clocks. A lot happened to time in the first twenty years of the twentieth century.

So Ruth Belville might be forgiven for feeling unsettled. Her personal experience of 1908 was a year of intrusion, anxiety and embarrassment, following St John Winne's lecture at the United Wards Club. Taking a step backwards, we see William Willett's scheme for changing the clocks being debated that year, and Winne's flagrant attempt to hijack the legislation for his own ends. Another step backwards gives an even wider view: Einstein had already published his special theory of relativity, which set the clock ticking for established concepts of time, and he was getting ready to publish his grand general theory. All of this was going on in 1908 and the years following. Germany, as we have seen, played a crucial role in both these wider movements, and so to Germany we can now briefly turn in Ruth's own story.

In January 1910, the Royal Observatory received a letter from a German journalist working in London, Marie Truandt. She had read an article in a German newspaper about Ruth Belville and wanted to interview her. Would the Observatory inform her of Miss Belville's address? The Observatory staff considered the matter and wondered what Ruth herself thought. 'As she comes every Monday I would suggest showing her the letter before a reply is given.' The letter was duly shown to Ruth the next time she made the journey from Ockwells Manor. She was not impressed. 'Miss Bellville', came the reply, 'has had so much correspondence on this subject that she is rather tired of it and would prefer that her address be _not_ given; but she leaves it to the Astronomer Royal to do as he thinks best.' A letter was sent to the journalist explaining Ruth's work, stating that 'it is understood that Miss Belville has certain clients whom she visits for the purpose of supplying them with the correct time, but we do not know who they are and whatever Miss Belville does in this way is quite unconnected with the Royal Observatory.' Ruth's address was refused to the journalist.

Soon after this unwelcome reminder to Ruth of her 1908 brush with the press, she decided that it was time to move house again. Sometime about 1911 or 1912, she made the journey on the Great Western Railway to London for the last time. She was now in her late fifties and her next stop was to be Ewell, a village in Surrey to the north of Epsom. Her new home was Ewell Cottage, a tiny dwelling on the London Road to the north of the village, a short walk from two railway stations with frequent services into London. She stayed in this house for over twenty years, sandwiched between two pubs, with her journey time to Greenwich cut by half.

So it was Ewell which was Ruth Belville's home during the First World War, but every Monday morning she would still take a train into London to change for the lines out to Greenwich, and make her usual visits, from the docks to the City, Chelsea to the West End, and then back out to Ewell Cottage. As German Zeppelins dropped their bombs on a defiant London, as German politicians set their clocks forward to *Sommerzeit*, as German scientists scratched their heads trying to understand Albert Einstein's universe-upsetting theories, Ruth Belville plied her route, regular as clockwork, her old pocket watch in her handbag, her black dress unruffled, her hat planted firmly on her head. Air raids took their toll in her customers' heartlands but Europe eventually woke up from its nightmare of war and, as the eclipse expeditions revealed Einstein's theories to a public desperate for a new world harmony, Ruth continued her work. She had seen off the Standard Time Company and the war; she had seen William Willett's summer time and Albert Einstein's relativity give her service new meanings. Now a new threat was waiting in the wings and this time change was quite literally in the air. For time had gone wireless.

Chapter 7

PIP! PIP! PIP! PIP! PIP! PIP!

{ 1920–1935 }

Dame Nellie Melba at the Marconi Works, Chelmsford, 1920.

NEW STREET, CHELMSFORD, TUESDAY 15 JUNE 1920. The
Australian soprano, Dame Nellie Melba, stood in front of a
makeshift microphone in the Marconi company's radio works in
Chelmsford. Around Britain and Europe, amateur radio enthusiasts
were glued to their receiving apparatus, headphones planted on
their heads, waiting to hear the first advertised British broadcast of
live public entertainment. At ten minutes past seven in the evening,
the Marconi broadcast, sponsored by the *Daily Mail* newspaper,
started to crackle across the airwaves. 'Hallo ... hallo ... hallo ...' An
announcer apologised to the 'listeners-in' for any reception difficulties caused by atmospheric conditions, but people in the south of
England ignored the apology; they could hear as clearly as if they
were standing in the studio at Chelmsford. Tweaks were made to
sensitive receivers; cats' whiskers were adjusted over delicate crys-
tals. Then Dame Nellie emitted a long trill, her own warning signal
that she was about to begin. Clocks ticked, breath was held, and

then the appointed hour arrived: quarter past seven, precisely. 'Home, sweet home': the famous soprano sang her trademark song and followed it with 'Nymphes et Sylvains' and the 'Addio' from Puccini's *La Bohème*. History was in the making; a live broadcast was in progress and the listeners-in felt as though they were in the Royal Albert Hall, not their living rooms. Melba knew her audience and did not leave them wanting. 'Chant Venétien' came next, followed by a repeat of 'Nymphes et Sylvains' before a rousing finale, the first verse of the National Anthem.

It nearly didn't happen. Melba was not the easiest artist to work with, and when she was shown the towering radio antenna masts atop the Marconi buildings and was told that it was from these that her voice would be carried around the world, she is said to have retorted, 'Young man, if you think I am going to climb up there you are greatly mistaken.' But the resulting broadcast was a triumph; its reception at the Eiffel Tower radio station was so strong it was reported that a gramophone record was made of her voice there. That night, Europe shrank to the size of a single room as, for the first time, one voice in Britain reached out to every living room in Europe within a thousand-mile radius. It was live, in real time, and could be heard by anybody with a radio receiver and an antenna. That night, in Chelmsford and in countless living rooms around Europe, radio broadcasting hit the airwaves.

Wireless communication was first successfully developed in the 1890s, with Guglielmo Marconi being credited with its invention although he was by no means the only person experimenting with radio waves at the time. By 1896, Marconi was working in Britain and had secured the first patent for radio. Its earliest use is evident in its first description: wireless telegraphy. It was originally a new means to fulfil the old function of point-to-point communication. The sending of messages from one person to another, originally using electric cables could now, with the advent of wireless, use empty space itself as the communication medium. Like the cable telegraph, early wireless message traffic consisted of the high-speed dots and dashes of Morse code, spelling out letters and symbols using a pattern of long and short bursts of sound or electricity. Radio waves enabled high-speed telecommunication without wires, without the complicated and expensive and essentially fragile infrastructure between sender and receiver that had been necessary for the functioning of the cable telegraph network. It was a tall order to maintain the forest of wires and poles and insulators and switches that loomed over city streets and extended along railway lines, postal routes and under the seas. How different it was that space itself might act as carrier; not in any way simple or technologically straightforward, but ripe with possibility.

The wireless opened up two distinct arenas of development. The first was the chance to communicate with ships at sea, sending messages, time and weather signals. The first regular experimental wireless time and weather service came from the US Navy in Washington DC and Boston from about 1904. Safe naval navigation relied on the setting of shipboard chronometers accurately to a standard (Greenwich) time, and where previously this had normally been done in port, with the new wireless technology ships had ready access to the time while at sea, enabling them to keep their chronometers more accurately to time and their navigation consequently more precise. Weather reports, too, aided safety at sea, as storms and other bad weather could be avoided or at least prepared for. In Europe, wireless time signals came a little later and they did not come first from Greenwich. Frank Hope-Jones, whom we met in the last chapter pontificating about Willett's summer-time scheme, was ready with a trenchant opinion on this matter too. In 1923, he wrote, 'The average Englishman imagines that Greenwich time comes from Greenwich . . . It used to, but unfortunately this country did not take the lead, did not take a fair share, nor, in fact, any share at all, in the establishment of the International Service of Wireless Time Signals.' The 'average Englishman' was able to pick up time signals but not British ones, because he 'has been getting it from the observatories and countries of his neighbours to an increasing extent for the last ten years.'

One can understand Hope-Jones's frustration. The neighbouring countries that had inaugurated wireless time signals in Europe were France and Germany, and it was the Eiffel Tower in Paris that formed the transmitter mast for Europe's most popular radio time service, formally set up in 1910. Without the radio, the Eiffel Tower could well have been demolished. It was built for the 1889 Paris International Exhibition and only on a twenty-year lease. But times had changed, and its capabilities as a transmitter mast were very clear by the turn of the century, so it was retained. The year 1908, as we have seen repeatedly throughout this account, was a turning point in the history of timekeeping, and the same applied in the story of wireless. In that year, three men in France were pushing to establish a time empire within the Eiffel Tower's lattice of struts and girders. Gustave Eiffel, the tower's builder, had already joined forces with Gustave-August Ferrié, a senior military engineer, to get the tower accepted officially as a radio station. Henri Poincaré, one of France's top scientists and intellectuals, joined the two men in 1908 to propose that the new station send out radio time signals for the benefit of shipping. They were pushing at an open door. The potential military and civil benefits from a wireless time service were clear and well understood, and by 1909 experimental transmissions had begun. They were a resounding success.

France, bitter that Greenwich had achieved the status of 'prime meridian' at the 1884 International Meridian Conference over the claims of Paris, saw in the Eiffel Tower the opportunity to preside over the greatest time service in the world. By July 1909, funds had been allocated to set up a permanent, regular service, and in May 1910, Paris Mean Time was made available to anyone with equipment to receive the signals: no wires, no subscriptions (apart from the cost of a licence), no weekly visits. The following year, France swallowed its pride, shifted its timescale by the nine minutes and twenty-one seconds that separated the Greenwich and Paris meridians, and thus Greenwich time from Paris was born. The equipment, though, was the sticking point. Eiffel Tower time was for ships' captains and professional astronomers, wireless enthusiasts and watchmakers. It was for specialists, not for Hope-Jones's 'average Englishmen'. Back in Britain, the outbreak of war in 1914 killed the market for the specialized crystal and magnetic radio receivers designed to pick up the French time signals. Civilian use of radio receivers was forbidden and it was not until 1919, after the war was over, that hobbyists could buy up military surplus components and construct equipment capable of picking up the distant coded signals. Ready-made receivers off-the-shelf were available but for the average Briton, what was there to listen to?

The second arena of possibility opened up by the wireless was broadcast entertainment – speech and music – alongside the dots and dashes of the specialist time services and the point-to-point message traffic. This was something different. This was about the content and form of the broadcasts, and it had more in common with newspapers than cable telegraphs. In the same way that one newspaper could spread its message over an entire nation, the radio brought everyone together in time and space in a way the cables had never been able to do, enabling shared moments to become national. Work on broadcasting started before the war but was interrupted by it because military requirements took priority. After the war, however, broadcast radio was taken up in earnest and Marconi's own works, as we have seen, hosted Britain's first advertised live broadcast in 1920 as Dame Nellie Melba belted out her crowd-pleasing numbers. Now, all those average Englishmen and Englishwomen had something to listen to besides dots and dashes.

Events moved swiftly on. The British Government tried to restrict growth in radio broadcasting, concerned that it might lose control of a lucrative and powerful new medium, and worried that civilian broadcasts might squeeze vital military radio transmissions out of the airwaves. But hobbyists, their appetites whetted by Dame Melba's trilling and thrilling arias, demanded a better service, and in 1922, after extensive lobbying by the radio manufacturing companies, the Government relented and allowed a group of leading wireless firms led by Marconi to join together as the British Broadcasting Company. For the first seven months of its existence, the nascent BBC broadcast in London from Marconi's own premises on the Strand, but the young firm was ambitious and soon outgrew its parent's accommodation, moving to rooms at the back of the Savoy Place headquarters of the Institution of Electrical Engineers. The official start of BBC broadcasting from Savoy Hill, as the premises became known, took place on 1 May 1923. A few days previously, however, our old friend Frank Hope-Jones had taken to the airwaves from Marconi House. Hope-Jones's topic was highly appropriate: he was there to talk about time.

Listeners to BBC radio on Saturday 21 April 1923, just before ten o'clock in the evening, heard one of the final BBC broadcasts from Marconi House, and it was an important one. In four hours' time, the clocks were due to be changed to British Summer Time, as they had done every spring since 1916 when William Willett's daylight-saving scheme was first put into operation. Hope-Jones, who had been instrumental in Willett's campaign, was about to give the BBC's listeners a summer-time reminder. But this programme did more than that. It revealed the future of the Greenwich time service: a five-second radio countdown. Hope-Jones explained to his audience that he was going to give them a time signal at exactly ten o'clock. He urged them therefore to set a watch forward one hour to 11 o'clock, to be started exactly when he gave the signal. The appointed hour approached. 'Stand by – are you ready?' He was about to count down the last five seconds of the hour and the sixth second would be the exact hour. 'One – two – three – four – five – six.' The watch was now set; the householder was instructed to go around the house, setting every clock and watch forward to eleven p.m., before retiring for the night. Ten days later, the BBC moved its studio facilities from Marconi House to the new accommodation in Savoy Hill. Hope-Jones had noticed that restricted space was not the only reason they needed to move: 'I was in the old studio speaking into a microphone supported with tin-tacks and string supported on an old soap-box – the station being apparently designed and made by W. Heath Robinson.' The microphone was not all that seemed out of step with the times. Beside it had stood a four-foot-high triangular frame supporting a set of eight tubular bells – the BBC's first time signal.

From the start, the BBC had informed listeners of the time. Sometimes the engineer (who acted as announcer) simply read it out; sometimes a gong was struck. The accuracy would often have been all right, because Marconi House had been fitted out before the war with a high-quality electric clock system. But the clocks were not always kept right, it seems, so sometimes the accuracy of the signals left a lot to be desired. And there was nothing systematic about it, no schedule to warn listeners when a time signal was due. Then there were those tubular bells, installed in Marconi House around December 1922. They were tuned to the famous Westminster chimes, and were played by the musical director, Stanton Jefferies, or the programme director, Arthur Burrows, to indicate the hour. Sometimes Jefferies played the chimes wrongly, and was inundated with letters and telephone calls of complaint. But he had a back-up plan. The Marconi House studios were just three-quarters of a mile along the river from the great Westminster clock and, if the wind was blowing from the south, the sound of Big Ben could easily be heard by opening the studio window. It was then an easy matter to play along on the tubular bells, but without making adjustments for the speed of sound the BBC's signal would go out about three seconds late. It was not ideal.

But things were looking up. On 5 April 1923, Godfrey Isaacs of Marconi had written to John Reith at the BBC to say that 'The "Daily Mail" have suggested to me that a very large number of people throughout the country very rarely have the right time, and it would be a very good idea if, at the end of our programme every evening, the Broadcasting Company announced the correct

Greenwich time.' Reith had already considered the matter. 'As soon as we move into the new studio,' he replied the next day, 'which should be in about three weeks time, we have an arrangement with the Standard Time Company to fit up a clock. The difficulty in Marconi House, I understand, is that the synchronised clocks are frequently wrong.'

The Standard Time Company again! The firm had clearly not lost the ability to push its services where it believed the future lay. The tubular bells moved with the BBC to Savoy Hill and the new studios, with their STC hourly signal, were also provided with a radio receiver that picked up the Eiffel Tower time signals, a wall clock with an easy-to-read second hand set daily to the French signals, and a portable chronometer that could be used as a back-up. At a stroke, the BBC time service was given a boost. Now, the announcer would alert listeners that the time signal – a single strike of the bell – would be given in one minute's time. 'Stand by, please for one minute, when I will give you the BBC's time signal for 7 o'clock.' Following Hope-Jones's April broadcast, they also counted out the five seconds leading up to the signal. But this was the BBC, and the new-born broadcasting service was ever keen to stay ahead of the time. Later that year, three key figures got their heads together to work out a more robust and standardized solution to the BBC's time signal problem. The BBC general manager, John Reith, wrote to the Astronomer Royal, Frank Dyson, requesting a meeting. Lurking behind the scenes was Frank Hope-Jones. Together, they decided that it was time to connect the BBC's broadcast transmitters to the Royal Observatory's clocks.

The Marconi company had considered the idea of a Greenwich automatic time signal as early as November 1922, the day after the newly formed BBC began its broadcasts from the roof of Marconi House. A letter from Marconi's research department to Arthur Burrows, its programme director, suggested 'that a time signal sent out periodically from the broadcasting station would be of great value to the public. Such a signal can be transmitted by the wireless set under direct control of Greenwich Observatory, and could be transmitted exactly at each hour.' This memorandum, which at the time led merely to the installation of STC clocks at Savoy Hill, had in fact been right on target. The trick was to connect the Royal Observatory's clocks directly to the transmitter, with no human intervention along the way, no spoken countdown, no tubular bells or Westminster chimes or hand-held chronometer. Just a clock, a wire, a buzzer and a transmitter; a direct, clear chain from the stars above Greenwich to the radio receiver in the home: Greenwich time broadcast directly to the country. On Friday 9 November 1923, at noon, Frank Dyson and his colleague William Bowyer called at 2 Savoy Hill, London, to see John Reith. The meeting went well in the end, although, according to Dyson's daughter and biographer, it began a little frostily. Reith opened by suggesting that public time broadcasts might enhance the Observatory's prestige but Dyson retorted that 'the prestige of Greenwich Observatory rested securely on its scientific record, and he neither sought nor wanted publicity.' Nevertheless, he did believe strongly in the ideals of public service, so he was happy to support the scheme.

Whilst the idea of an automatic time signal had been mooted by several people and then agreed in principle at the November meeting, the form of the signal, it seems, was proposed by Hope-Jones. Dyson and Hope-Jones knew each other quite well. In 1923, the Astronomer Royal was planning to have a new type of pendulum clock installed at the Observatory, called the 'free pendulum', designed by William Hamilton Shortt and manufactured by Hope-Jones's Synchronome company. These super-accurate clocks were taking the observatory world by storm and soon, with Hope-Jones's involvement, Dyson replaced all the main Greenwich timekeepers with Shortt's new device. It was therefore with his colleagues back at Greenwich, and his energetic collaborator Hope-Jones, that Dyson discussed the practical details of an automatic time signal following his meeting with Reith. Hope-Jones put forward a scheme involving five short 'dots' of sound, akin to the familiar dots of Morse code used in wireless telegraphy and existing radio time signals, but here spaced apart by one-second intervals, like his spoken five-second count broadcast the previous year. Dyson modified the proposal by adding a sixth dot to act as the actual time signal itself following the five-second warning – creating, as his successor noted, 'Sir Frank Dyson's signature tune.' In fact, Hope-Jones was actually recycling a practice that already existed. According to Herbert Jones, a Synchronome employee for over fifty years, the firm ran a roaring trade with local clockmakers by counting out the last five seconds of any minute down the telephone, according to their factory clock set right to the Eiffel Tower signals every day. STC provided a similar service.

With the principles agreed and the practical form of the signal decided upon, Dyson set rapidly to work, and days later, he wrote to Reith saying that he had arranged for two spare Observatory clocks to be fitted with electrical contacts by the noted chronometer-makers Victor Kullberg at a cost (to the BBC) of £20 per clock. These contacts enabled the ticking of the clock's pendulum and rotation of its gear-wheels to send an electrical time signal to the BBC every thirty minutes; it was later changed to every fifteen minutes, to fit better with programme schedules. By January, the work had been done and the half-hourly signals, consisting of six short bursts of electricity, were shooting along a rented telephone line from Greenwich to Savoy Hill, ready to be turned into audible tones. The BBC was now awash with time: the Eiffel Tower, daily; the Standard Time Company, hourly; Greenwich time every thirty minutes. Time was marching onwards, and the date was set for a grand inaugural broadcast of the new British time service. On 5 February 1924 at 9.15 p.m., Frank Dyson addressed the nation from Savoy Hill, giving a short lecture on the history of Greenwich timekeeping. And then, at the conclusion of his speech fifteen minutes later, the first automatic Greenwich time signal was broadcast.

In all the excitement of this new, modern service, we should not miss the fact that the BBC actually inaugurated two precision time signals that month. Alongside the six dots of the Greenwich Time Signal (known to BBC insiders, then as now, as the GTS), the BBC started regular broadcasts of Big Ben. The Royal Observatory and Big Ben had a long history together. The Victorian Astronomer Royal, George Airy, was heavily involved in specifying the clock's design, and a special telegraph wire had been laid from the clock tower to the Observatory so that Airy and his successors could keep track of the clock's timekeeping accuracy. Big Ben began to tick in 1859 and by 1924 the grand old timekeeper at Westminster had more than earned its spurs, rarely being more than a second wrong. It was a prime candidate, then, to act as one of the BBC's two new time signals – a role it plays to this day.

Big Ben's first appearance on the wireless was to usher in the new year of 1924. It was something of a surprise item in the night's listening schedules. The *Radio Times* had told listeners to expect to hear *Auld Lang Syne* at midnight before close-down five minutes later, but its director, John Reith, had something a little more exciting in mind. He wanted to re-create the feeling of live celebration; the sentimental and atmospheric experience of an outdoors party, a ball. Music was fine as far as it went, but that New Year's Eve in 1923 Reith wanted to liven things up a little, and what more sentimental way to toast the new year than the sound of the country's favourite clock? In an early example of an 'outside broadcast', BBC engineers rigged up a microphone and amplifier on the roof of St Stephen's Club, a building on the junction of Bridge Street and Victoria Embankment, directly opposite the clock tower housing Big Ben. This was no easy feat in those early days of radio: Bridge Street was over half-a-mile from Savoy Hill, requiring a substantial run of cabling, although the routing was simple, straight up the Victoria Embankment. At the stroke of midnight, the sound of Big Ben rang across the air of Westminster and into the transmitters of the BBC, ringing in a very special new year to listeners around the country.

The experiment was a success, though traffic noise and weather conditions would cause problems, so a better solution was needed if the great clock was to provide a daily BBC time signal. A second line was therefore run up into Big Ben's lofty enclosure, alongside the Greenwich time telegraph line from the Observatory. The new cable connected the Savoy Hill studios with a microphone suspended in the clock tower near the bells. The microphone was packed into a wad of cotton wool which was in turn sealed into a rubber football bladder to keep the weather out. This low-technology approach was all that seemed necessary for this simple system, although an early engineer's inspection following the microphone's installation found that weather was not the only problem in the draughty clock tower. Birds had pecked through the rubber bladder and made off with most of the cotton wool. It was fixed and with this permanent wiring-up the famous clock bell began its regular live transmissions at three o'clock on the afternoon of Sunday 17 February 1924. With two new national time services on offer, timetables were published in the press telling listeners when to expect the GTS and when Big Ben could be heard. A couple of months later, Frank Hope-Jones clarified matters by informing listeners that Big Ben was usually right to within a second or two whereas the GTS was accurate to better than one-fiftieth of a second. He said the BBC was 'transmitting two kinds of time music, low-brow and high-brow . . . Big Ben is good enough for the large majority . . . we do not want to trouble low-brows with a code.'

Speaking of codes, here's a riddle: when is a pip not a pip? When it's a dot. We know the BBC's famous time signal today as 'pips', but that was not always the case. For the first few years of its existence, it was the 'six-dot' signal. It makes sense. The time signal grew out of practices employed in the days before the BBC, before broadcast radio, when the wireless was used for sending text messages and coded time reports. This was a world of dots and dashes, so it is only natural that the terminology would carry forward to the new broadcast world. Hope-Jones was ready with a full explanation. To readers of the *Radio Times* in 1925, he said, 'it was thought desirable to give the public something very simple, and preferably something which they had become accustomed to . . . I considered dots better than dashes for accurate clock comparisons.'

But when did Britain decide that the six staccato 1000-hertz bursts of tone sounded more like 'pips' than 'dots'? Pips were all the rage in those days. In 1918, after the end of the First World War, British politician Eric Geddes had said that 'the Germans . . . are going to pay every penny; they are going to be squeezed as a lemon is squeezed – until the pips squeak.' Pips were spoken to imitate bicycle or motorcar horns, leading to P. G. Wodehouse's famous 1920s cheery 'pip-pip!' But austere BBC time signals? It's impossible to be sure when that all started, but no reference to 'pips' has been found so far before about 1930. Then comes the year 1932, notable in our story for two reasons. Firstly (and Ruth Belville probably didn't know this), on 5 May St John Winne died at the age of 70 in a Weston-Super-Mare hotel. Then just nine days later, Frank Hope-Jones made yet another of his broadcasts on the BBC which just might be the first recorded utterance of the now-famous time-signal description, 'pip'.

It was Saturday night, and the BBC was having a moving-out party. The following day was to be the official launch of its new headquarters building, Broadcasting House, north of London's busy Oxford Street, and it was a much-needed move. The BBC had long since grown out of its rented accommodation in Savoy Hill, and to celebrate the move to the purpose-built flagship building, the BBC's final broadcast from its old home was a retrospective programme entitled 'The End of Savoy Hill'. The show started at twenty past nine in the evening and ran until midnight. The problem with a retrospective programme, though, was that the BBC had only gained the capacity to record their broadcasts in the previous couple of years, using the new magnetic steel tape-recording instrument called the Blattnerphone (after its inventor, Ludwig Blattner). So the 'clips' show that night in 1932 was a mixture of recently recorded shows and re-readings of programmes that had gone out years before but had never been recorded. The BBC Year Book for 1933 picks up the story. 'Those who heard "The End of Savoy Hill" in May may be interested to know that as many as nineteen different programmes in the course of the evening appeared in "flashes" from the Blattnerphone.' One man who was never far from new technology – or the chance of self-publicity – was the ubiquitous Frank Hope-Jones, who was about to re-read to the nation his 1923 broadcast about Willett's daylight saving time, at the end of which he had counted out the last five seconds before ten o'clock. Here he was in 1932, trying to remember what he had said nine years before. We can still listen-in to the recording, as the announcer made the introduction: 'Here is Mr F. Hope-Jones, our Time Expert.'

'Since the introduction of Daylight Saving in May 1916, the public have had to rely upon the newspapers to remind them to alter their clocks. Thanks to radio-telephony and the British Broadcasting Company . . . that reminder can now be given at the appropriate moment, just before you perform that time-honoured domestic rite of winding-up the clocks on Saturday night.' But the clock was ticking, and it was soon time for Hope-Jones to give his listeners a very special time signal. 'Now take out your watch. Put it forward exactly one hour with great care. The time at this moment is very nearly ten, so you must turn the hands to eleven. You have just time to do this before I give you the exact time signal at the hour. I shall do it by counting out the last five seconds of the hour.' He coughed, excitedly. 'Stand by, are you ready? The last five seconds of the hour, and the last one will be the exact hour. Pip! Pip! Pip! Pip! Pip! Pip! Now go around the house and move the minute hand of each clock one revolution forward, allowing it time to strike . . .'

Pip! That now-familiar sound, introducing the news of the day, the six short bursts that cry it's time – time to set your watch, time to listen, time at Greenwich. He said the word 'pip' that evening in 1932, but had he really counted out pips, not seconds, back in 1923? Almost certainly not. We do know how he counted down his summer-time reminder in 1924, in April, after the automatic GTS had started, and if people were calling the dots 'pips' then, he would surely have used the word in his own spoken broadcast, but he didn't. He counted, '1, 2, 3, 4, 5, 6.' In his 1932 broadcast for the end of Savoy Hill it's quite likely he was being slightly economical with the truth. In the 1920s, the pips were simply dots. The programme came to an end at midnight. One minute later, Savoy Hill broadcast its last ever 'good-night', and a few seconds after that, a voice was heard over the airwaves: 'Broadcasting House calling. That is the end of Savoy Hill . . .' The end of an extraordinary decade.

At the same time that the BBC was establishing its time signals, Ruth Belville, the old-time Greenwich time service, was investing in her own technology. By the 1920s, her pocket chronometer by John Arnold was over 130 years old and, while it was kept in good order by the Royal Observatory's technical staff, Ruth clearly felt she needed a back-up watch in case her Arnold needed to go in for an extended service – or in case he was damaged or stolen. Ruth was already thinking of her own mortality. She made a will in 1921. Then, in May 1926, Britain was rocked by a ten-day general strike. London's transport networks virtually came to a halt and at the bustling docks, scenes of soldiers attempting to keep order in the face of near-riot would have been alarming, to say the least, to anybody trying to get about with a bulky silver-cased pocket watch. So, on 21 July 1926, Ruth Belville visited the Royal Observatory to collect a large pocket chronometer by Charles Frodsham, bought by the Admiralty in 1890 for £18. The very next day, she bought it – at a price unknown – and Charles Frodsham number 07586 was struck off the books of the Navy. Charles had joined Arnold, and the new boy was a star. In Observatory trials in early 1890, the watch had come fourth in a class of twenty-four tested for accuracy and stability – a fine result.

Charles had been sent all over the world in the service of the Navy, appearing as far afield as Gibraltar and Sydney. In the Great War he served on the Royal Fleet Auxiliary ship *Ferol*, which was an oil tanker, and then on a couple of submarine depot ships, HMS *Lucia* and HMS *Vulcan*. These acted as mobile 'mother ships' for fleets of diesel submarines, providing engineering support, weaponry, and accommodation and facilities for the submariners where there were no fixed naval bases. But after the war was over, the watch was hardly used, and the Navy was clearly happy to see it put to better purpose outside the military world. Given its record of accuracy it was a great choice and while there is no firm evidence the general strike prompted Ruth to make the purchase, she must have felt more secure knowing that Charles could take over if anything happened to Arnold. She described the watch as 'gallant' and once said to a customer, 'I've brought Charles today since Arnold is not very well.'

While the general strike of 1926 reminded Londoners of the potential power of mass action, it also focused attention on aspects of mass communication: in particular, the ownership of the broadcast media. Here a state project was being realized in two notable ways. Firstly, the British Broadcasting Company was brought into public ownership in 1927 and renamed the British Broadcasting

Corporation. Secondly, a major new wireless communications facility was opened in the Warwickshire countryside, called Rugby Radio Station. This 900-acre Post Office facility at Hillmorton was first proposed in the early 1920s as a means to communicate via high-powered radio with the British Empire overseas. The station opened officially on 1 January 1926 and, for the first time, British radio broadcasts could be heard on the other side of the world. The station used long-wave radio transmissions from an aerial slung on twelve 820-foot masts, each mast assembly weighing over 200 tons. This was a high-power service. A current of 700 amperes flowed in the huge aerial, creating waves that could follow the curve of the Earth rather than radiating off into space. So, as well as providing the facility to send and receive messages anywhere in the world, the Rugby station was ideally placed to radiate time signals that could be picked up by ships overseas to ensure safe navigation. Connections were made between Rugby and the clocks at the Royal Observatory, and on Monday 19 December 1927, the first time signals were broadcast. Greenwich time had gone truly global. Rugby Radio Station broadcast Greenwich time for eighty years until 2007, when the service moved to Cumbria. The gigantic aerial masts were brought down with explosive charges soon after.

So, at the end of a remarkable decade, 1930 saw a new time landscape for Britain. Radio receiver ownership had risen dramatically and listeners could now hear the BBC's six pips and the striking of Big Ben every day in their living rooms. Clockmakers and wireless enthusiasts could pick up specialist coded signals from the Eiffel Tower, the Rugby Radio Station and other international transmitters broadcasting time and weather reports. Frank Hope-Jones, both by himself and through his Synchronome company, had raised the profile of accurate timekeeping enormously, and the Astronomer Royal, Frank Dyson, had responded energetically, boosting the Observatory's timekeeping technology and connecting it to the world. The Standard Time Company continued to send its hourly Greenwich time pulses along its London wire network, and the Royal Observatory, through the General Post Office, regulated Britain's railways, post offices and communications networks with its own brand of electrical timekeeping. William Willett's summer time scheme was also put permanently on the statute book.

But amid all this modernity, all this precision, all these grand schemes, did anybody still have time for Ruth Belville, still resolutely journeying between Ewell Cottage and the docks, watchmakers, shopkeepers and factories of London? Was time running out for the Greenwich time lady? Not a bit. As we have seen, Ruth invested in her own technology, which might seem humble in comparison with the pips and dots and wireless broadcasts, but pocket chronometers remained the finest portable timekeepers around, and her service worked, and was reliable. New technology doesn't just sweep aside old systems. They co-exist for far longer than one might expect. Even in the 1930s, for instance, market workers in east London received a daily time signal from knockers-up blowing dried peas through pea-shooters at their bedroom windows. Whatever worked remained appropriate. Another example: the Post Office didn't just broadcast the time from Rugby or telegraph it over the wire network. In 1931 they announced that new Venner time switches installed in their telephone kiosks, for switching on and off the night-light, could be used by the public as a quick time-check, as the devices had a tiny clock dial visible through a window. There was no linear succession here, just layers of complexity.

So Ruth still had a place, although the radio rattled her. 'The wireless is my greatest enemy today,' she told a newspaper in 1929. But you didn't need a licence to receive Ruth's weekly visit. You didn't need to buy any gadget, maintain any equipment, repair any electrical connections or worry about the wind on your wires or the rain on your aerial. You simply paid your 'sub' and got your weekly knock at the door. The only commitment was the provision of tea and a few words of small talk. Londoners had not finished with Ruth Belville, and let us also recall that Britain was going through a great economic depression in the 1920s. Times were tight and if those visits by Ruth Belville weren't needed and valued by her customers, then they would not have continued paying her annual fee. The fact that people continued to pay for Ruth is evidence enough that her service was still valuable, despite the other options opening up that decade. But by the early 1930s, Ruth was in her late seventies, and she had no heirs. How much longer could she continue? Who could possibly replace the Greenwich time lady?

Chapter 8

AT THE THIRD STROKE ...

{ 1935–1939 }

Ethel Cain on a cigarette card, 1935.

GENERAL POST OFFICE HQ, CITY OF LONDON, FRIDAY 21 JUNE 1935. A young telephone operator who worked for the Victoria Telephone Exchange, one of London's largest, was reading Milton into her microphone. 'And at my window bid good morrow, through the sweet-briar, or the vine, or the twisted eglantine.' But this was no ordinary operator call. Ethel May Cain, 26, was hidden away in Room 12 of the GPO's Headquarters Building in St Martins-le-Grand, a couple of streets north of St Paul's Cathedral, and she was reading seventeenth-century poetry to a group of literary and critical luminaries listening in a hall nearby. 'What do you think of that voice, Miss Thorndike?' Stuart Hibberd, the chief announcer of the BBC, had a gilt telephone receiver pressed to his ear, listening to Ethel's rendition of Milton's 'L'Allegro'. Dame Sybil Thorndike, sitting by his side, replied, 'I think it's a charming voice. Beautiful tone ... sense of rhythm ... I should like to hear it with time sentences.' Stuart Hibberd agreed, an instruction was relayed by telephone to Ethel in her reading room, and so began the voice that launched a thousand lunch-hours. 'At the third stroke, it will be ...'

Ethel Cain was a finalist in the GPO's first ever 'golden voice' competition. To judge it, the GPO recruited not only Stuart Hibberd and Sybil Thorndike, but also the Poet Laureate, John Masefield (who designed the speaking tests), and Lord Iliffe, the newspaper magnate, representing commerce. Their task? To select the perfect telephone voice from 15,000 exchange operators across England, ready to record it for the latest in timekeeping technology: the telephone speaking clock. For months, operators had taken part in competitive heats to find the best voice. Their trade union had considered matters and finally decided it would be a good thing for the girls, and by June 1935 the contestants had been whittled down to nine. And here they were, Ethel and her colleagues, reading poetry to the judges.

There was one further judge sitting in the hall with the poet and the announcer and the actress and the businessman, and she was having a very special day out. While the telephone operators were competing to be 'the girl with the golden voice', they were also acting as judges in their own competition: to find the perfect telephone subscriber from all the customers who called exchanges to get their calls connected. 'In every telephone exchange certain subscribers become known to the telephonists for the helpful manner in which they transact their calls,' explained *The Times,* and these favoured customers were nominated up and down the country to find the final member of the judging panel. But like the golden voice competition itself, only women were involved, as *The Times* observed: 'the attitude of a male subscriber to a female telephonist not being deemed as always due to telephonic considerations'. Mrs Atkinson, of Burley-in-Wharfedale, Leeds, was the winner. The judging panel was complete and Rita Atkinson was whisked down to London, all expenses paid, first class. 'The perfect telephone subscriber meets the perfect telephone operator,' said a report.

The judges sat around a billiard table covered with a cloth in the GPO's King George V Hall, opened just three weeks before by the Prince of Wales. All of them listened intently to their gilt telephone receivers (stock reference: 'old gold tele.162') specially installed for the occasion to judge the nine finalists selected from all those 15,000 woman telephonists across the country. The master of ceremonies was Major Tryon, the new Postmaster General. In the background, filming the contestants reading their lines and the judges discussing their diction, were Pathé Gazette newsreel camera operators, capturing the historic scenes forever. The contestants read their sublime poetry and prosaic time sentences in Room 12 and the judges listened to their receivers in the hall. Throat lozenges were passed around the contestants by the supervising matron and glasses of lubricating water gulped down.

Thanks to the cameras and the news reporters and the weeks of auditions, the ears of the nation were listening to Miss Cain and her colleagues reading Milton and the time of day, for this was a remarkable moment in timekeeping history: 'It is a great office to be the very voice of Time speaking to an inquiring and uncertain world, to strike the golden notes that will command the silvery chimes,' an editorial in *The Times* told readers. Britain's telephone speaking clock, known as TIM (short for TIME, the number to call on the alphanumeric telephone dials of the day), went on to become one of the most popular and well-used time-distribution systems ever built, and it all started in that hall, with those judges, listening to poetry through gilt telephones around a hastily-disguised billiard table while discussing the charms of young women's voices. 'They demanded from the competitors a voice that was beautiful in quality, having fullness of tone, with nothing niggardly about it, and nothing rasping in the breathing or in the note,' explained the *Post Office Magazine* in August 1935. Ethel was announced the winner.

At the end of the day, prizes were handed out to all the finalists. Miss Dunn, of the London Trunk Exchange, took silver to Ethel Cain's gold, and the other seven took joint bronze, one of these runners-up being Mary Dixon from the South Shields Exchange. The Poet Laureate spoke to a reporter: 'All the voices had qualities of beauty and interest, and only in one were we able to find any trace of accent, and that was a slight Northern one.' That would be Miss Dixon. Better luck next time. The judges enjoyed a celebratory luncheon of salmon, lamb cutlets in aspic and gooseberry fool, over which the Postmaster-General presided. Then everyone trooped off to the Prince of Wales Theatre, just off Leicester Square, where, according to a Post Office report, Miss Cain and Mrs Atkinson 'addressed the audience'. That summer, the Prince of Wales theatre was showing 'La Revue Splendide', a non-stop French revue in English. That afternoon, when Ethel Cain and Rita Atkinson came on their official visit, the audience that they 'addressed' had come to see a cast including the 'Glorious Girls', 'the Boy Friends' and 'the Show Ladies', singing musical numbers such as 'We'll Show You A Thing Or Two'. These shows were for businessmen with a couple of hours on their hands and 'an appreciative eye for the female form', as one reviewer put it. The question must be asked. What *exactly* did the Post Office have in mind when it sent the 26-year-old, tall, slim, blonde Ethel Cain to the Prince of Wales that afternoon to 'address the audience'? Unfortunately, no record of what she said – or wore – seem to have survived, which is a shame.

On 4 July, Ethel recorded the 79 different phrases that make up the speaking clock onto large glass discs, the sound being stored photographically in the same way that movie soundtracks were recorded at the edge of the filmstrip. The recordings took a while. She had to get the diction just right for the very limited frequency response of the telephone microphones and receivers of the day. 'Thirty' and 'forty' could sound the same, unless she forced the 'forty' forward and sent the 'thirty' very far back, and for years afterwards, Post Office engineers received complaints that the clock was ten minutes (or ten seconds) wrong because people had misheard the words. Nevertheless, Sybil Thorndike had told Ethel Cain after the competition that she had perfect diction. She was certainly used to talking on the telephone. Ethel and the other Victoria Telephone Exchange operators handled some 375,000 calls per week. Victoria wasn't converted to automatic dialling until 1939, so in 1935 all those calls were connected by Ethel and the girls. You want to make a call? You pick up the handset and wait to speak to an operator. 'Engaged. Line engaged.' Selfridges took out advertisements in *The Times* to ask, 'How very frequently have you, who use the telephone hour after hour, found yourself attracted or repelled by the voice that answers you!' Milton must have been a breeze.

Just over a year later, on 24 July 1936, TIM the talking clock went live from a pair of machines in the Holborn exchange. Again, Major Tryon presided over the ceremonies, while the job of cutting the ribbon – or rather, the ceremonial lifting of a red telephone handset

– fell to the new Astronomer Royal, Harold Spencer Jones. John Pond had inaugurated the Greenwich time service in 1833 with the time ball, because he was sick of London shop lads knocking at his door when he was trying to sleep, asking to know the time. In 1836, George Airy had commanded his assistant, John Belville, to send the time down to the watchmakers directly. A century later, in exactly the same fashion, Major Tryon was switching on the automated speaking clock because he was fed up with subscribers calling his switchboard girls to find out the time by the exchange clock – 100,000 Londoners each month were tying up valuable lines and operators and switches and time by telephoning the exchange for a time-check. 'The time by the exchange clock is ...' Prescribed wording; no mention of 'the correct time', just in case the exchange clock was wrong.

Be active, not passive, thought Tryon, as Airy had done exactly 100 years before. Send the time out, don't let the people come knocking, and better still, charge for the service. Build a clock that talks, connect it by wire to the Royal Observatory's hourly time service to keep it right, push 200 telephone lines into its innards and set it running. Leave the exchange girls free to connect real calls. The red telephone was lifted, the glass discs ran up to speed, Ethel Cain's disembodied golden voice lilted from its seventy-nine individual movie soundtracks, and there it stood: TIM, the talking clock, telling perfect Greenwich time to anyone who would listen, all day, every day, never getting tired, never making a mistake, never rasping, taking 200 callers at once and giving each one a beautifully enunciated time-check with three precise strokes.

The calls started coming thick and fast. The GPO traffic engineers (an elite band, with their own magazine, called, not surprisingly, *Traffic*) were ready for an avalanche of what they termed 'curiosity traffic', by installing extra operators to connect calls from subscribers without direct dialling, by installing extra lines, extra switches, even allocating a special number which only the operators knew to call if the main lines into TIM were busy ('The number . . . must not be disclosed to members of the public in any circumstances'). An avalanche indeed it was. Just in its first year of operation, TIM took nearly 20,000,000 calls, and that was when it was only available in London. Twenty million London time-checks to Ethel Cain in 1936–7, compared with Ruth Belville's 2,500. There was a new Greenwich time lady and she was a beauty. 'Miss Cain has one of the most beautiful voices I have ever heard, and behind a beautiful voice you will also find intelligence,' explained the Poet Laureate after the golden voice competition final. Only the best girls were good enough for the GPO's telephone exchanges. Here's a circular memorandum from the GPO Secretary, dated 17 July 1901, on rules for the selection of telephone operators: 'The age at the time of admission should be between 17 and 19 years. The minimum height should be 5 feet 2 inches without boots or shoes.' The switchboards were quite tall constructions. It would have been a problem if operators couldn't reach the top. Then assuming they passed the practical tests, they were all given a voice examination. Naturally, good hearing and speech were important. But the rules went further. 'Any candidates showing any indication of nervousness, hysteria, want of self-possession, or a strongly marked local dialect should be rejected.' Still in the running? You'd better be bright, too. 'Candidates will be required to pass an educational test . . . it is therefore desirable that well-educated girls should be selected.'

Major Tryon and the rest of the golden voice competition judges therefore had their pick of some top-quality exchange operators for the golden voice competition. Perhaps Mrs Atkinson, the perfect telephone subscriber, had been brought in to act as chaperone during the competition finals. The *Post Office Magazine* writer certainly took something of a shine to the golden girl. 'Miss Cain, who is a blue-eyed, slim blonde in the mid-twenties, has a beautifully modulated voice, calm and unflurried.' What a voice! What a figure! Beautiful *and* intelligent! And remember, twenty million time-checks in the first year alone. A remarkable success for the nascent service, worthy of handsome reward to the voice behind the clock, surely. One might have thought so, but Ethel's payment for this ground-breaking piece of timekeeping history was just ten pounds and ten shillings. One ten-thousandth of a penny per call to the speaking clock, just in the first year. Ten guineas to buy a girl with a golden voice. In later years she thought it was fool's gold.

Ethel Cain became a 'celebrity' – but in fact she was not the first speaking clock. As so often happens, new ideas challenge old ideas, but don't necessarily turn them over. Back in 1916, Britain was at war, and with the impending clocks change, Ruth Belville's customers wanted access to the right time more frequently than ever before. So they devised a plan. On 28 April that year, a letter from Ruth's little cottage in Ewell was delivered to the Astronomer Royal, Frank Dyson: 'Sir, it has been brought to my notice by some of the Watch & Chronometer Makers that it would be a great convenience to them at present if I could supplement my weekly visits by giving them G.M. Time by Telephone one day weekly.' Ruth, the first telephone speaking clock? Not quite. Her nemesis, the Standard Time Company, had offered a service (for a guinea a year) for the right 'to ring us up as often as you please for the exact time'. This was in 1912. But Ruth's customers clearly preferred to hear the time lady, not the company man. She had the technology. Her customers had offered her an office in the City of London with a telephone and a very good pendulum clock, and she would call into this office with her Arnold, just as she would to any other weekly subscriber, but here she would just find an empty chair and she would start telephoning the chronometer-makers who did not have the spare staff capacity to find the time any other way. There was, after all, a war on, and these people were engaged in vital work for the Admiralty, servicing the chronometers and watches that kept Britain's navy afloat (or rather, kept it in the right place). Ruth and Arnold could help, but did Frank Dyson mind? 'Miss Belville informed verbally that she is entitled to take the time from the Clock. What use she makes of it must be her own responsibility.' So there was no problem there, and three weeks later all the country's clocks changed. Everyone needed the time. The lady with the watch telephoned the time from her London office with her pocket watch and her very good pendulum clock and it was twenty years before we heard of her usurper, Ethel Cain, with her golden voice.

So Ruth was there in the telephone time world too, twenty years before the Post Office built their automatic machine called TIM who spoke with Ethel's voice and told the time directly from the Greenwich Observatory. Even Frank Hope-Jones and the Standard Time Company had been telephoning the time to *their* customers, long before TIM. Forward again to that time, to 1936, to the new girl speaking to a world that had been changing its clocks twice each year for two decades because it seemed like a good thing to do. What did this all mean for Ruth Belville?

Frank Thirkell used to work for Mappin & Webb, the Royal silversmith and famous chain of jewellery and silver stores. He remembered meeting Ruth on her weekly visits to his store in the 1930s, when she was far into her eighties. He first met her in 1936 at the Regent Street showrooms, and then in 1939 he transferred to the main Oxford Street store and she called there too, still bringing the time once every week from Greenwich until it was time for her to retire, at long last, probably in 1940, as the bombs were dropping once more and her time was almost up. Ruth knew the clock was ticking and she knew who was responsible and she didn't mind telling Frank Thirkell one time she was visiting Mappin & Webb. 'TIM. What a vulgar name, don't you think?' She asked him that, but we don't know what he said in reply. We do know, however, that he later left the employment of Mappin & Webb and became a director of Charles Frodsham, the high-class chronometer firm that had provided Ruth's understudy watch back in the 1920s, so he must have been impressed with the service provided by Ruth's mechanical time signal. TIM was always there, always willing and eager but, like the telegraph time service, sometimes got it wrong: the Post Office blamed an error of two minutes, in December 1936, on a faulty telephone line from the Royal Observatory. TIM just told you the time, over and over. Nobody had time for Ruth any more. Everybody had fallen in love with young TIM. 'In view of the possibility of certain members of the public becoming so enamoured of the golden voice that they are impelled to listen to it for an indefinite period, an automatic device disconnects the circuit at the end of three minutes.'

Time was a small world and it seems it was a world full of fate. Ruth Belville had been on the move since she was born. She started out in Greenwich, and then the Belville women moved out to the new-builds of Charlton. Four Charlton properties later Ruth moved out to Maidenhead, then onwards to Ewell. She liked it there, and stayed a long time. In the early 1930s, though, a thundering bypass road was built right by her cottage. The properties next to hers – the houses of Ernest Pidgeon and Ralph Thomas, and the 'Organ Inn' – were demolished. In their place, a grand new public house, the 'Organ and Dragon', was built, a typical new Tudor-style pub with a big car park, the sort of facility springing up on bypass roads across Britain. With the noise of the road, the cars parking up for their beer and sandwiches directly next to her cottage, the disturbance in her life, Ruth had to move again. By 1934, the now octogenarian spinster was living in what was to be her final residence, a yellow-brick cottage in Plough Lane, Beddington. This was a short tram ride along the Stafford Road to the nearest railway station, Waddon, one stop away from West Croydon, and on a direct line to London Bridge and its trains out to Greenwich.

A year later, the newspapers were full of Ethel Cain, the new Greenwich time lady, selected from thousands of exchange operators from around the country. It could have been anybody but it was Ethel, living with her parents in the modest family home at 108 Mitcham Road, West Croydon, which happened to be a twenty-five-minute walk from Ruth Belville's house, or five minutes on the tram. Of all the women in all the country's telephone exchanges, the new time lady chosen in 1935 lived round the corner from Ruth Belville. As Ruth's train from Waddon stopped at West Croydon, she might have unknowingly seen Ethel waiting for a London Victoria train to take her to work at the telephone exchange. Ruth Belville and Ethel Cain were practically neighbours.

At Croydon's local history archive, in the town's central library, there is a file building up on the telephone heroine. It even reveals that Ethel actually worked in the library before she joined the Post Office. But across the borough boundary into Sutton, where Ruth Belville's house was located, the equivalent local archive contains no such file, no such press clippings, no memory of the Greenwich time lady. She was a latecomer, after all; a Greenwich girl who ended up in Plough Lane only at the end of her long life. For the West Croydon local communities, Ruth was the new girl; Ethel the old-timer. Gilded publicity for Croydon's golden girl; no blaze of glory for Ruth's twilight years.

By 1939 and the outbreak of the Second World War, Ruth was getting ready (if not quite willing) to pass on the baton of Greenwich time lady to Ethel May Cain. Ethel was a keen actress. Since she was a teenager, she had played in several dramatic clubs in Croydon, and once she got the speaking clock job she must have known she was a star. Stars aren't called Ethel May, however, and she speedily changed her name to Jane Cain, one worthy of a film star, and kept it forever more, or at least while she shot her first 'picture', the 1935 Columbia film entitled *Vanity*. As the *Croydon Times* reported of her rise to fame, 'Always with a smile in her eyes. Jane Cain, one-time telephonist, likes being a film star.'

Ruth, too, had changed her name on taking over her mother's time business, since she was really Elizabeth Ruth Belville, but had used her middle name from childhood. Sometime about 1890 (she would have been in her thirties) she again started using Elizabeth in censuses and official documents, though she remained Ruth in local directories and personal correspondence. Her father, John Belville, had changed his name to John Henry on taking up his time job, though he perhaps had more reason to do so, given the French terror.

With all of this change in the 1930s, Ruth was feeling her own time catching up with her. Three years after TIM went live, Britain was at war once more. Just as John Belville had survived the Franco-British wars at the turn of the nineteenth century, so Ruth had continued working through the Great War with its air raids targeting her own customers. But in the Second World War she was not to be so lucky. She was about to begin her final chapter.

Chapter 9

HUMAN 'TIM' FOUND DEAD

{ 1939-1943 }

Thomas Lewes Sayer in The Sketch, *17 July 1935.*

57 PLOUGH LANE, BEDDINGTON, 7 DECEMBER 1943. A call was made to the local police station. PC Flower was despatched to Ruth Belville's house in Plough Lane, one of a small row of pretty brick cottages near the railway line to Croydon. When he got through the front door, he noted a strong smell of gas in the house. He quickly moved upstairs to the bedroom, where he found the body of an old woman in bed. Beside the bed, he reported, lay an old silver pocket watch. Ruth Belville, the Greenwich time lady, aged 89, was dead.

Although she was old and unmarried, Ruth was not without friendship and support in her later years. Her mother's extended family, based in Norfolk, still kept an eye on her. One of her third cousins, Frances Maria Read (born Sayer), was about the same age, and the two were lifelong friends. Frances had been born the year after Ruth, but died a year earlier, aged 87, in October 1942. One of Frances's grandsons, Bryan C. Read, recalls taking tea with Ruth in the 1930s during her trips to Norfolk to visit Frances. The family was always fascinated by Ruth's unusual occupation and her pocket watch, Arnold, was legendary. Another third cousin was Thomas Lewes Sayer, who lived a few streets away from Ruth, in Hall Lane, Wallington. Sayer was a prominent figure in the City of London, having worked for forty-nine years in the Town Clerk's office before retiring in 1930. He had seen a lot in his time. This was the man who, in 1888, had viewed the mutilated remains of Catherine Eddowes, one of the Ripper victims. He saw action in the First World War, and on 30 March 1920, he was made a Member of the Order of the British Empire for 'services in connection with the War'. After his retirement, he became a Surrey County councillor and wrote a book about his experiences in the City Corporation. Attending funerals and writing obituaries had formed a significant part of his official life, one of tradition, duty and loyalty. He was eleven years younger than Ruth, married with several children, and he kept a close eye on her in her last years. It was probably to live near the Sayers that Ruth sensibly moved to Plough Lane in the first place.

It was Sayer who kept Ruth's memory alive after her death, and who told the world of the services she had rendered to horology over her long life. And it had been Sayer, as we shall see presently, who had used his City connections to secure Ruth a decent income in her retirement. Six days after her death, Sayer attended the coroner's inquest in nearby Sutton, which concluded that the cause was not old age but carbon monoxide poisoning, brought about by fumes from a faulty gas lamp. There was no ventilation in the room, according to the police officer who found the body, and Charles Baker, an inspector for the Croydon Gas Company, also gave evidence. 'We warn people, when we clean these lamps, not to turn them down low,' he said. The gas lamp was badly sooted up. 'This had been turned low for a considerable time, and, because of the soot, which had formed in the nozzle, there was a certain amount of carbon monoxide.' The coroner for Surrey, Cyril Baron, probed deeper: 'Are these sorts of lamps cleaned regularly?' Baker replied that this was the case in peacetime but not in wartime. Then they were only cleaned on request.

Ruth Belville had retired from the time supply business a couple of years previously, after the outbreak of the Second World War. We do not know exactly when but it was probably in the autumn of 1940. She was a spinster, with no heirs to take over the family business. She had already worked her way through one war, then the second one came. Her home was directly north of Croydon Airport, a major target for bombing raids, night after night, and it was no better at the other end of her journey. Always vulnerable given its proximity to the river and the docks, on the morning of 16 October 1940, the Royal Observatory took a direct bomb hit on the main gates to the front courtyard. Every window in the institution was shattered, the gates were destroyed, and the white dial of the clock that had been telling Londoners the time since 1852 was blown out of its housing. The blast forced the 36-inch-diameter copper disc inwards against the mechanism and then sucked it outwards into the shape of a dish, and it wasn't replaced until 1947. Miraculously, though, the bomb just missed the transit telescope and there was no structural damage to any of the buildings. It was a lucky escape.

A reserve time station had already been set up outside London, in Abinger, Surrey. After the bombing of autumn 1940, most of the Observatory's time department decamped to this safe house, which became the transmitting station for Greenwich time signals for the remainder of the war. Greenwich was no longer the home of time, and London was no place for an octogenarian to be walking the streets. The city was becoming deranged. Roads were shattered, whole blocks destroyed. Every morning, Ruth would not know which of her customers had survived, which premises were still standing. But the real killer for the Belville time-supply business had been TIM, the talking clock. Who wanted a weekly visit from Ruth when Ethel Cain was at the end of the telephone, all day, every day? It was a natural end for the business, and time to call it quits. Ruth retired, spending more and more time in the Beddington cottage where she had lived since 1934, until she fell victim to the silent killer, carbon monoxide.

The gas company engineers reminded customers not to turn down their lamps too low, or they would block up with soot, but it was no surprise that an old lady forgot to call them in to clean the lamps, and without regular cleaning there was no reminder. It became a vicious circle, as noted by the coroner: 'I doubt if it is generally known that there is any risk in turning an incandescent gas mantle to a low position, especially in these days when people are encouraged to keep their gas consumption as low as possible.' He then recorded a verdict of accidental death. Ruth, unlike the Royal Observatory, had had no lucky escape. Like everyone else, she was doing her bit for the war effort by conserving precious stocks of gas. She survived Zeppelins and the Blitz, but in the end the war claimed her as well. The *News of the World* reported her passing: 'Human "T.I.M." Found Dead.'

Sayer took care of the arrangements. He placed a death notice in *The Times*, and on the day of the inquest, the newspaper ran a short obituary he had written, explaining how Ruth had 'devoted half a century to taking the correct Greenwich time to business houses in London on a watch 100 years old.' Next, he contacted a local funeral director, Truelove, to organize a cremation. Home Office regulations on cremation had been relaxed at the outbreak of war to ease matters following air raids, and by 1945 the cremation rate had doubled to nearly 8 per cent. On 15 December 1943, eight days after her death and two after the inquest, Elizabeth Ruth Belville was cremated at the South London Crematorium, Streatham. This private facility had been built in 1936 as part of the Streatham Park Cemetery that had opened in 1908. Both had been momentous years for Ruth Belville. The service was conducted by a minister named King, and afterwards, Ruth's ashes were scattered in the crematorium garden. The register records the very spot where this final act of memory took place: plot AW51, then, as now, in a peaceful lawned garden next to the buildings. Sayer arranged for an entry to be made in the book of remembrance, filling a whole page with beautifully flowing cursive script: 'In memory of Elizabeth Ruth Belville (Horologist). Died 7th December 1943 aged 89 years.' It is still on display there.

It might have been the end of the Belvilles but Arnold, the pocket watch, lived on. Ruth always wanted him to have a good home once she had gone. She knew how valuable he was: not financially, but as an icon, a symbol of time distribution in the days before radio, before telegraphs, before TIM the talking clock. 'When it retires from business, which I hope will be a long time yet, I think the dear old thing ought to be put in the British Museum,' she had told the *Evening News* in 1908. The journalist from the *Maidenhead Advertiser* recommended the Maidenhead Museum. Yet neither of those museums received Arnold. In 1908, Ruth was only in her fifties but times changed. Thirteen years later, aged 67, she decided to make her last will and testament, and money was on her mind. 'All my just debts and funeral and testamentary expenses shall be paid as soon as conveniently may be after my decease . . . I should wish my funeral to be as inexpensive as possible.'

At that time Ruth was hardly living in idle luxury. Apart from a few sticks of furniture, a rented cottage and some pictures and china, all she really had was her business, a pocket watch or two, her own clocks, and her subscriber book. If she died in service, debts from subscribers would need to be called in, which would take time. Arnold could help. 'Should my death occur while carrying on my appointment of supplying G.M. Time from the Royal Observatory my chronometer Arnold 485 also my clocks could be left as security to Messrs. Stephen Tree & Co . . . who would grant an instalment of money upon them until payments due to me are called in.' Stephen Tree was almost certainly Ruth's 'Borough' customer that she mentioned to the *Kentish Mercury*, but somewhat ironically, back in the 1880s, the Tree company in Great Dover Street was a subscriber to the Standard Time Company's wired time service whose propaganda campaign caused Ruth such distress. Tree was prepared to help if necessary, however, but he would have to give up Arnold when the time came. 'I should wish the Royal Observatory to have the first offer of purchasing my chronometer,' Ruth had written in her will, but the Royal Observatory was no more successful than the British Museum in securing Arnold when the time came, and Tree's business would not be able to help either.

By May 1941, with Hitler's bombs raining down on England, over a million buildings had been destroyed or damaged. The previous month, a type of Luftwaffe high-explosive bomb, weighing nearly two tonnes and nicknamed 'Satan' by British bomb disposal men, had first been dropped on London. The location was Great Dover Street, creating, as one eye-witness described it, a 'gigantic sheet of flame'. The blast caused damage over a huge area, levelling all the properties nearby including Stephen Tree's clockmaking premises at numbers 120–122. He would not be able to help after all.

By then Ruth Belville had retired from the time business, in her eighties, and needed a pension to see out her last days. Fortunately the community that she and her parents had served so faithfully for so long was ready to step in: the watchmakers and chronometer companies who had needed her most and had prompted the formation of the time-carrying business in the first place. It was their trade guild, the Worshipful Company of Clockmakers, who saw to it that Ruth Belville would be able to set Arnold down by her bed for the last time. In January 1941, a vacancy arose on the Company's pension list, and Ruth was elected to fill it. In April 1941, the Company's court (their managing committee) approved the election and allocated £15 per year, to be made in quarterly payments starting immediately.

In October 1942, Thomas Lewes Sayer wrote to the Company on Ruth's behalf asking if the pension might be increased. A further vacancy had arisen and the court agreed to raise her pension to £30 per year. Sayer, the City insider, had helped keep his old friend solvent. In the same meeting, the court discussed a vacancy that had arisen on the court of assistants (ordinary members of the committee) and the Astronomer Royal's name, Harold Spencer Jones, was put forward. Spencer Jones was the man who had continued his predecessors' policy of welcoming Ruth Belville each week to his observatory and had provided her with tea while her watch was checked. He was also the man who had lifted the red telephone handset in 1936 and made the first call to TIM, the talking clock, which had such a *vulgar* name and a voice straight from Croydon.

In January 1943, Ruth wrote to the clerk of the Clockmakers' Company. When the time came, would they like Arnold for their museum? The court discussed the offer and replied that they would be pleased to find a home for the watch. 'By doing so her name would be perpetuated for all time,' the meeting minutes observed. In April 1943 (the clock is ticking), Harold Spencer Jones attended his first court meeting and eight months later the court met again and the clerk reported the death of Miss Elizabeth Ruth Belville. The master directed that Truelove's funeral bill of £25 be paid by the Company. The clerk had already been handed Ruth's pocket watch, presumably by Sayer, and Arnold had made the journey from the Belville home to the City watchmakers for the very last time. After the war was over, the museum reopened, and there sits pocket chronometer number 485 by John Arnold to this day, not at the Royal Observatory, nor the British Museum, nor in Maidenhead. Arnold went where he truly belonged, to the watchmakers in London, and he could be forgiven for wanting a rest. The watch was made in 1794. Ruth died (and the watch, therefore, ran down) in 1943. That's 149 years, or to put it another way, over 4.7 thousand million seconds. The watch beats five times each second, so the number of ticks it had performed in its long, eventful life is a figure that is hard to imagine. Arnold had earned his keep.

That just left the rest of Ruth's small estate. Her will named Lewes Hector Read, one of Frances Maria Read's sons, as her executor, but the will, made in 1921, was now badly out of date. One of Lewes Hector's sisters, Mary May, was Ruth's sole benefactor, but she had died in 1934. Who was next in line? Another of the Read siblings, Margaret Ethel, had married a Norwich solicitor, Leonard Hill, who advertised for claimants of Ruth's £200 estate to come forward, but what became of the search remains a mystery.

This was truly the end of the line for the Greenwich time ladies but the Second World War had also almost spelled the end for Ruth's erstwhile rival, too. Just a month after the Clockmakers' Company started making pension payments to Ruth Belville, the Luftwaffe mounted the largest bombing raid seen on London during the war. Among the 500 aircraft launched on Saturday 10 May 1941, one released a total of four incendiaries between London Bridge and Southwark Bridge. The fires spread quickly and fiercely, and soon reached the heart of Queen Victoria Street. The pendulum clocks, electrical relays, wires and batteries pulsing out the hourly time signals at the Standard Time Company were about to tick for the last time; the heart of the company was about to stop beating. The block of buildings that included numbers 19 and 21 Queen Victoria Street, STC headquarters since 1882, was caught up in the conflagration and was swiftly gutted. One of the most chilling images of London at war shows the frontage of the block in mid-collapse as fire-fighters struggled to contain the blaze. Ruth Belville's chronometer had outlasted all of St John Winne's clocks.

Perhaps surprisingly, though, the 1941 destruction of STC's headquarters building was not the end of the company's service. Hundreds of miles of overhead telegraph lines had been destroyed along with the building, but thanks to the energy and efforts of one of its longest-serving employees, A. B. Webber, the company found new offices in nearby Old Broad Street. It built new control equipment, strung up replacement wires, and got a replacement time service up and running very quickly, albeit serving a much smaller area of London.

The STC time supply business in fact survived until the 1960s, another instance of the recurring theme in this story of Greenwich time, that new ways did not stop or replace old ones, but simply added to them and brought further choice. Today, the BBC six pips still sound on the radio, as does Big Ben. The speaking clock is getting more calls than ever before. Hourly time signals along London telegraph wires stopped in the 1960s, but the network is still there; modified, refurbished, now carrying the digital high-frequency broadband signals of the internet with the time servers that regulate your computer clock, but in many cases still entering the home or office as a simple pair of copper wires. The more things change, the more they stay the same. Even the practice of transporting time by carrying clocks didn't end with the death of Ruth Belville, although the aim became rather more elevated. In the 1970s, a series of experiments saw Einstein's theories of relativity tested by carrying super-accurate atomic clocks aboard aircraft, finding out whether changes of speed and gravity really did make clocks run at different speeds. They did. Einstein was right.

It is easy to assume that Ruth Belville's hand-carried service was an anachronism by the twentieth century, that it was merely, as the *Evening News* put it in 1929, 'a wonderful old lady's business trips with a wonderful old watch'. In 1975, a historian of the Royal Observatory, A. J. Meadows, wrote briefly about the Belvilles' weekly time service but was briskly dismissive. 'In the later years, these visits were maintained more as a tradition than out of necessity: much more accurate methods of obtaining time-signals were by then readily available . . . Indeed, by the time of [John] Henry's death Airy had already started on a much better scheme for distributing time.' But this view is a distorted one. New technologies do not simply sweep away older ones, especially when the old one continues to work as well or better for a particular purpose. Of this Ruth's chronometer is a celebrated example, keeping time as well at the end of its long service (and after 24 thousand million ticks) as it did when made in 1794, and often more reliably and accurately than more complex electrical systems. The crucial point, however, does not relate to the technology so much as its users. All systems involve people, and people are complicated and make decisions about technology for complicated reasons. To understand the history of technology we need to consider human psychology and the complex process of how and why people change their behaviour.

From the users' perspectives, Airy's telegraph time service, and that of St John Winne's Standard Time Company, was good but not always good; available but not always readily so; accurate enough for most people most of the time but no more so than Ruth's service, which was always reliably correct to a tenth of a second and thereby continued to retain a loyal specialized following for as long as she had the endurance to continue. New isn't necessarily better, it's just different, and the solution people go for depends on what they need.

Stuff endures. Getting Greenwich time from a lady with an excellent pocket watch in the 1920s or 1930s was just as valid and just as appropriate as getting it from the wireless or the telegraph lines. In many ways it was a more robust system, more trustworthy. Ruth turned up. You didn't have to call out an engineer to fix Ruth. You knew Ruth's signal was right because you could see the handwriting on the weekly certificate from Greenwich and you knew how well the watch was performing compared to every other time you'd seen a certificate.

Ruth also provided what no electrical wire could: the personal touch. People are inherently social animals, and in commercial relationships as in any others, good service creates long-lasting human bonds that are difficult to break even when new technological alternatives are available. The Belville service eventually became an institution, but that is not why it lasted so long. It lasted because it worked, because Ruth was good at what she did, because what she did was necessary for her customers. But in the end, even the Greenwich time lady could not stop the passage of time.

SOURCES AND FURTHER READING

General sources used throughout all chapters

Census returns.
Registers of electors.
Births, marriages and deaths registers.
Local street directories.
Contemporary and modern street maps of Greenwich, Charlton, London, Maidenhead, Ewell, Croydon, Beddington.
The Times.
Ralph Hyde, ed., *The A to Z of Victorian London,* Harry Margary and the Guildhall Library (London, 1987).
Ann Saunders, ed., *The A to Z of Edwardian London,* London Topographical Society (London, 2007).
Oxford Dictionary of National Biography, Oxford University Press (Oxford, 2004); online edition, hereafter 'Oxford DNB'.

Archives and libraries consulted

BBC Written Archives Centre, Reading.
Berkshire Record Office, Reading.
Bishopsgate Institute, London.
British Library, London.
British Library Newspaper Collection, Colindale.
British Pathe online archive (www.britishpathe.com).
Bromley Central Library and Archives.
BT Archives, London.
Croydon Local Studies Library and Archives.

Ewell Public Library and Local History Centre.
Charles Frodsham & Co. Ltd., London.
Greenwich Heritage Centre, Woolwich.
Guildhall Library, London.
HM Courts Service, Principal Probate Registry, London.
Institution of Engineering and Technology Library and Archives, London.
Lewisham Local History Centre.
London School of Economics Library.
Maidenhead Central Library.
The National Archives, Kew.
National Maritime Museum Library and Archive, Greenwich.
National Meteorological Archive, Devon Record Office, Exeter.
Royal Greenwich Observatory Archive, Cambridge University Library, Cambridge.
Science Museum Library, London.
Sutton Local History Centre.

Chapter 1: 1795–1856

National Meteorological Archive, Exeter: John Henry Belville, 'Meteorological Observations'.
Royal Greenwich Observatory Archive, Cambridge: George Airy papers, RGO 6/3, 'Papers on Government Superintendence, 1849–1854'.
Royal Greenwich Observatory Archive, Cambridge: William Christie papers, RGO 7/96, 'Correspondence on Chronometers, 1895–1923'.

Royal Greenwich Observatory Archive, Cambridge: Philbert Melotte papers, RGO 74/6/2, 'Account of J. H. Belville and GMT'.

Marriage certificate, John Henry Belville and Maria Elizabeth Last.

Birth certificate, Elizabeth Ruth Naomi Belville.

Family papers, 1947–1960, private collection.

Personal communication, Bryan C. Read, son of Lewes Hector Read, 2006 and 2008.

Thomas Earnshaw, *Longitude: An Appeal to the Public ...* , F. Wingrave (London, 1808).

John Henry Belville, *A Manual of the Barometer*, Richard and John Edward Taylor (London, 1849).

John Henry Belville, *A Manual of the Thermometer*, Richard and John Edward Taylor (London, 1850).

Edwin Dunkin, *A Far Off Vision: a Cornishman at Greenwich Observatory*, Royal Institution of Cornwall (1999) (autobiography originally written 1894).

Autobiography of Sir George Biddell Airy, ed. Wilfrid Airy, Cambridge University Press (Cambridge, 1896).

Vaudrey Mercer, *John Arnold and Son, Chronometer Makers 1762–1843*, Antiquarian Horological Society (London, 1972).

R. H. G. Thomas, *London's First Railway: the London & Greenwich*, B. T. Batsford Ltd. (London, 1972).

Christian Wolmar, *The Subterranean Railway*, Atlantic Books (London, 2004).

Nautical Magazine, 28 October 1833 and vol. IV, 1835.

Greenwich Observations and *Astronomer Royal's Reports*, Royal Observatory (Greenwich), annually, 1836–1857.

Illustrated London News, 9 November 1844.

Illustrated Exhibitor, 8 November 1851.

Kentish Mercury, 19 July 1856.

'Plan of the Buildings and Grounds of the Royal Observatory, Greenwich, 1863, August; with Explanation and History', appendix II, *Greenwich Observations*, Royal Observatory (Greenwich, 1862).

The Observatory, vol. 20, 1897.

William Nash, 'Monthly Rainfall at the Royal Observatory, Greenwich, 1815–1903', *Quarterly Journal of the Royal Meteorological Society*, vol. XXX, no. 132, October 1904.

Philip Laurie, 'The Greenwich Time Ball', *The Observatory*, June 1958.

John Hunt, 'James Glaisher FRS (1809–1903): Astronomer, Meteorologist and Pioneer of Weather Forecasting: "A Venturesome Victorian"', *Quarterly Journal of the Royal Astronomical Society*, vol. 37, 1996.

John Hunt, 'The Handlers of Time: the Belville Family and the Royal Observatory, 1811–1939', *Astronomy & Geophysics*, February 1999.

Oxford DNB, 'John Pond', C. Andrew Murray.

Chapter 2: 1856–1892

Guildhall Library Manuscripts Section, London: Parkinson & Frodsham papers, 'Register of Clocks Watches etc 1866–1914', MS 19909.

Lewisham Local History Centre, London: St Margaret's Church, Lee, burial register, 1838–1861.

National Meteorological Archive, Exeter: John Henry Belville papers, 'Greenwich Temperatures, 1811–1856'.

Royal Greenwich Observatory Archive, Cambridge: George Airy papers, RGO 6/4, 'Papers on Government Superintendence, 1855–1860'.

Royal Greenwich Observatory Archive, Cambridge: George Airy papers, RGO 6/25, 'Astronomer Royal's Journal, 1848–1861'.

Royal Greenwich Observatory Archive, Cambridge: George Airy papers, RGO 6/30, 'First Assistant's Correspondence, 1855–1860'.

Royal Greenwich Observatory Archive, Cambridge: George Airy papers, RGO 6/43, 'Papers Concerning Occasional Observers, 1849–1873'.

Royal Greenwich Observatory Archive, Cambridge: Philbert Melotte papers, RGO 74/6/2, 'Account of J. H. Belville and GMT'.

'Bouët-Willaumez', *La Grande Encyclopédie*, H. Lamirault et Cie (Paris, 1887).

Autobiography of Sir George Biddell Airy, ed. Wilfrid Airy, Cambridge University Press (Cambridge, 1896).

T. Lewes Sayer, *Gog and Magog and I: Some Recollections of 49 Years at Guildhall*, Sampson Low, Marston & Co Ltd (London, c.1931).

Margaret Wilson, *Ninth Astronomer Royal: the Life of Frank Watson Dyson*, Heffer & Sons (Cambridge, 1951).

The Streets of London: the Booth Notebooks – South East, ed. Jess Steele, Deptford Forum Publishing (London, 1997).

Kentish Mercury, 19 July 1856.

Astronomer Royal's Reports, Royal Observatory (Greenwich), read 6 June 1857.

David Rooney, 'Maria and Ruth Belville: Competition for Greenwich Time Supply', *Antiquarian Horology*, September 2006.

David Rooney and James Nye, 'Greenwich Observatory Time for the Public Benefit: Standard Time and Victorian Networks of Regulation', *British Journal for the History of Science*, published online by CUP, 15 July 2008.

Oxford DNB, 'Sir James Whitley Deans Dundas', J. K. Laughton, revised by Andrew Lambert.

Oxford DNB, 'Cecilia Louisa Glaisher', Caroline Marten.

Oxford DNB, 'James Glaisher', H. P. Hollis, revised by J. Tucker.

Oxford DNB, 'William Mark Noble Mariott', J. Malcolm Walker.

Chapter 3: 1892–1908

Greenwich Council, London: Charlton Cemetery, burial register.

Royal Greenwich Observatory Archive, Cambridge: William Christie papers, RGO 7/58, 'Papers on Greenwich Park, 1876–1912'.

Royal Greenwich Observatory Archive, Cambridge: William Christie papers, RGO 7/253, 'Correspondence Regarding Time Signals, 1879–1893'.

Royal Greenwich Observatory Archive, Cambridge: William Christie papers, RGO 7/254, 'Correspondence on Time Signals, 1884–1913'.

Death certificate, Maria Elizabeth Belville.

Letter, family papers, 1959, private collection.

Personal communication, David Parr, railway clock specialist, 2008.

On Horological Telegraphy, Etc, John Flack & Co. (London, 1876).

International Conference Held at Washington for the Purpose of Fixing a Prime Meridian and a Universal Day, protocols of the proceedings (Washington DC, October 1884).

Karl Baedeker, *Baedeker's London and its Environs 1900,* facsimile, Old House Books (Moretonhampstead, 2002).

Joseph Conrad, *The Secret Agent* (1907).

Margaret Wilson, *Ninth Astronomer Royal: the Life of Frank Watson Dyson,* Heffer & Sons (Cambridge, 1951).

Ian Bartky, *One Time Fits All: the Campaigns for Global Uniformity,* Stanford University Press (Stanford, CA, 2007).

Daily Graphic, 31 October 1892 and 1 November 1892.

Daily News and Leader, 29 April 1913.

P. S. Laurie, 'Fireworks at the Royal Observatory', *Castle Review,* March 1955.

Mary T. Brück, 'Lady Computers at Greenwich in the Early 1890s', *Quarterly Journal of the Royal Astronomical Society,* 1995.

New York Times, 13 July 1996.

Mary T. Brück, 'Lady Computers', *Astronomy Now,* January 1998.

Mary Croarken, 'Mary Edwards: Computing for a Living in 18th-Century England', *IEEE Annals of the History of Computing,* October–December 2003.

David Rooney, 'Maria and Ruth Belville: Competition for Greenwich Time Supply', *Antiquarian Horology,* September 2006.

James Nye and David Rooney, 'Such Great Inventors as the Late Mr Lund: an Introduction to the Standard Time Company, 1870–1970', *Antiquarian Horology,* December 2007.

David Rooney and James Nye, 'Greenwich Observatory Time for the Public Benefit: Standard Time and Victorian Networks of Regulation', *British Journal for the History of Science,* published online by CUP, 15 July 2008.

Chapter 4: January–March 1908

Guildhall Library Manuscripts
Section, London: United Wards
Club papers, Transactions and
Committee Minutes, MS
21483/1, MS 11724/2 and MS
11723/1.

Institution of Engineering and
Technology Archive, London:
Silvanus P. Thompson papers,
file SPT 65/52, 'The Time of a
Great City', March 1908.

Royal Greenwich Observatory
Archive, Cambridge: William
Christie papers, RGO 7/96,
'Correspondence on
Chronometers, 1895–1923'.

Royal Greenwich Observatory
Archive, Cambridge: Philbert
Melotte papers, RGO 74/6/2,
'Account of J. H. Belville and
GMT'.

Vaudrey Mercer, *The Life and Letters
of Edward John Dent*, Antiquarian
Horological Society (London,
1977).

Alan Jackson, *London's Termini*, 2nd
edn., David & Charles (Newton
Abbot, 1985).

Horological Journal,
August–December 1904.

Maidenhead Advertiser, 11 March
1908.

James Nye and David Rooney, 'Such
Great Inventors as the Late Mr
Lund: an Introduction to the
Standard Time Company,
1870–1970', *Antiquarian
Horology*, December 2007.

Oxford DNB, 'Horatio William
Bottomley', A. J. A. Morris.

Chapter 5: March 1908

Getty Images Fox Photos Archives.

Royal Greenwich Observatory
Archive, Cambridge: William
Christie papers, RGO 7/96,
'Correspondence on
Chronometers, 1895–1923'.

Letter from G. W. Rickett (Royal
Greenwich Observatory
chronometer department),
family papers, 1959, private
collection.

Personal communication, John
Liffen, Science Museum, 2008.

Personal communication, David Parr,
railway clock specialist, 2008.

Karl Baedeker, *Baedeker's London and
its Environs 1900*, facsimile, Old
House Books
(Moretonhampstead, 2002).

*Bradshaw's General Railway and Steam
Navigation Guide for Great Britain
and Ireland*, April 1910, facsimile, David & Charles (Newton
Abbot, 1968).

Donald de Carle, *British Time*, Crosby
Lockwood & Son (London,
1947).

Vic Mitchell and Keith Smith, *East
London Line*, Middleton Press
(Midhurst, 1996).

Desmond Croome, *The Piccadilly
Line*, Capital Transport
(Harrow, 1998).

Alan Jackson, *London's Local
Railways*, 2nd edn., Capital
Transport (Middlesex, 1999).

David Leboff and Tim Demuth, *No
Need to Ask: Early Maps of
London's Underground Railways*,
Capital Transport (Harrow,
1999).

The Moving Metropolis: A History of
London's Transport Since 1800,
ed. Sheila Taylor, Laurence
King / London Transport
(London, 2001).
Docklands Light Railway Official
Handbook, 5th edn., Capital
Transport (Harrow, 2006).
Pictorial Times, 18 April 1846 (some
print runs only).
Daily Express, 7, 9 and 10 March 1908.
Maidenhead Advertiser, 11 March,
25 March and 22 April 1908.
Kentish Mercury, 13 March 1908.
Manchester Guardian, 8 May 1908.
The Observatory, July 1908.
Daily News and Leader, 29 April 1913.
The Observer, 24 and 31 August 1913.
Evening News, 3 April 1929.
Obituary, Elizabeth [Ruth] Belville,
The Observatory, vol. 65, 1944.
John Hunt, 'The Handlers of Time:
the Belville Family and the
Royal Observatory, 1811–1939',
Astronomy & Geophysics, February
1999.
First Great Western railway time-
tables, 2008.

Chapter 6: 1908–1920

National Maritime Museum
Archives: William Willett
correspondence, 1907–1914.
Royal Greenwich Observatory
Archive, Cambridge: William
Christie papers, RGO 7/96,
'Correspondence on
Chronometers, 1895–1923'.
Sir Edward A. Sassoon, A Lecture on
the Telegraph Lines of the Empire,
London Chamber of Commerce
(London, 1902).

Report and Special Report from the Select
Committee on the Daylight Saving
Bill together with the Proceedings of
the Committee, Minutes of Evidence,
and Appendix, HMSO (London,
1908).
Report and Special Report from the Select
Committee on the Daylight Saving
Bill together with the Proceedings of
the Committee, Minutes of Evidence,
and an Appendix, HMSO
(London, 1909).
The Ingenious Dr Franklin, ed. Nathan
Goodman, University of
Pennsylvania Press
(Philadelphia, 1931).
Gertrude Magrane, 'Recollections
of William Willett', in Petts
Wood 21st Birthday Festival
Week Souvenir Programme, 16–22
May 1948, Bromley Central
Library.
Isaac Asimov, Asimov's Biographical
Encyclopedia of Science and
Technology, David & Charles
(London, 1978).
Who Was Who, vol. I (1897–1915),
A & C Black (London, 1988).
Chambers Dictionary of Quotations,
Chambers (Edinburgh, 2005).
Ian Bartky, One Time Fits All: the
Campaigns for Global Uniformity,
Stanford University Press
(Stanford, CA, 2007).
Daily Express, 8 November 1919.
Obituary, Thomas Daniel Wright,
Horological Journal, December
1933.
A. O. Aldridge, 'Franklin's Essay
on Daylight Saving', American
Literature, vol. 28(1), 1956.
Notes and reading lists from a
Relativity Reading Group,

Cambridge University,
October–November 2003.
David Rooney, 'The Rightness of
Lightness: Contextualising
Daylight Saving and
Commemorations of William
Willett, 1918–1948', in
preparation.
Oxford DNB, 'Benjamin Franklin',
J. A. O. Leo Lemay.
Oxford DNB, 'William Willett',
Andrew Saint.

Chapter 7: 1920–1935

BBC Sound Archive, London:
'Extracts from a Talk by F.
Hope-Jones ...', in *The End
of Savoy Hill*, audio recording
and catalogue entry, transmis-
sion date 14 May 1932, written
and constructed by Lance
Sieveking.
BBC Written Archives, Caversham:
Greenwich Time Signal papers.
Charles Frodsham & Co. Ltd.,
London: company manufactur-
ing books.
National Maritime Museum Archive,
Greenwich: Admiralty
chronometer ledgers.
National Maritime Museum Archive,
Greenwich: Royal Observatory
watch trials results.
Royal Greenwich Observatory
Archive, Cambridge: Philbert
Melotte papers, RGO 74/6/2,
'Account of J. H. Belville and
GMT'.
Will and grant, St John Winne.
Personal communication, Mike
Todd, BBC Broadcast Duty
Manager, 2005.

Personal communication, Richard
Stenning, Director of Charles
Frodsham & Co. Ltd., 2007.
Personal communication, Cdr Peter
Linstead-Smith OBE, retired
submariner, 2008.
W. G. W. Mitchell, *Time & Weather
by Wireless*, The Wireless Press
(London, 1923).
A. O. Gibbon, *The Electrical Control
of Time Services in the British Post
Office*, Institution of Post Office
Electrical Engineers (London,
1930).
BBC Year Book, 1933.
Alfred Gillgrass, *The Book of Big Ben:
the Story of the Great Clock of
Westminster*, Herbert Joseph
(London, 1946).
Donald de Carle, *British Time*,
Crosby Lockwood (London,
1947).
Margaret Wilson, *Ninth Astronomer
Royal: the Life of Frank Watson
Dyson*, Heffer & Sons
(Cambridge, 1951).
*BBC Sound Recording: its Engineering
Development*, British Broadcasting
Corporation (London, 1962).
W. J. Baker, *A History of the Marconi
Company*, Methuen (London,
1970).
Edward Pawley, *BBC Engineering
1922–1972*, BBC Publications
(London, 1972).
Herbert Edward Jones, *The Lives of
Edith Dorothy Jones and Herbert
Edward Jones*, unpublished family
memoir, compiled 1978–9.
Vaudrey Mercer, *The Frodshams: the
Story of a Family of Chronometer
Makers*, Antiquarian Horological
Society (London, 1981).

Tim Wander, *2MT Writtle: the Birth of British Broadcasting*, Capella Publications (Stowmarket, 1988).

The Oxford Dictionary of Quotations, ed. Angela Partington, revised 4th edn., Oxford University Press (Oxford, 1996).

Tom Ridge, *Central Stepney History Walk*, Central Stepney Regeneration Board (London, 1998).

Ben Thomas, *Ben's Limehouse Recollections*, Ragged School Museum Trust (London, 1998).

Peter Galison, *Einstein's Clocks, Poincaré's Maps*, Hodder & Stoughton (London, 2003).

Horological Journal, May 1923 and May 1924.

Radio Times, issues covering December 1923 and January 1924; 11 September 1925.

E. H. Shaughnessy, 'The Rugby Radio Station of the British Post Office', Institution of Electrical Engineers Wireless Section, *Proceedings*, 1926.

'A Tour Round Savoy Hill', *Wireless World*, 9 March 1927.

Evening News, 3 April 1929.

Watch and Clock Maker, 15 November 1931.

Humphry Smith, 'International Time and Frequency Coordination', *Proceedings of the IEEE* [Institute of Electrical and Electronics Engineers], May 1972.

Geoffrey Goodship, 'Time in Broadcasting', *Electrical Horology Group Technical Papers*, no. 46, 1991.

David Read, 'The Transmission of Time Signals by Wireless', *Antiquarian Horology*, Spring 1998.

Geoffrey Goodship, 'Time and the BBC', *Antiquarian Horology*, Winter 1998.

Museum of the History of Science, Oxford, *Wireless World: Marconi and the Making of Radio*, special exhibition, 25 April to 1 October 2006.

Chapter 8: 1935–1939

BT Archives, London: subject file, 'Speaking Clock'.

BT Archives, London: subject file, 'Victoria T/E'.

BT Archives, London: file POST 33/4799.

BT Archives, London: image file, 'Recorded Services'.

BT Archives, London: sound recordings of Ethel Cain, Pat Simmons, Brian Cobby and Sara Mendes da Costa.

'The Golden Voice', 24 June 1935, Pathe Gazettes newsreel BP24063581943.

'Time Please!', 4 April 1938, Pathetones newsreel BP040438116625.

'Speaking Clock (aka Time)', 4 January 1954, New Pictorials newsreel BP040154133922.

Royal Greenwich Observatory Archive, Cambridge: William Christie papers, RGO 7/96, 'Correspondence on Chronometers, 1895–1923'.

V&A Theatre Collection Archive, London: scrapbook, 'Prince of Wales, 1935'.

Report and Special Report from the Select
Committee on the Daylight Saving
Bill together with the Proceedings of
the Committee, Minutes of Evidence,
and an Appendix, HMSO
(London, 1909).

Standard Time Co. Ltd., products
and services catalogue, 1912.

Donald de Carle, British Time, Crosby
Lockwood & Son (London,
1947).

GPO Research Report 20723: The
Post Office Speaking Clock Mark
III, GPO (London, 25 March
1964).

Vaudrey Mercer, John Arnold & Son,
Chronometer Makers 1762–1843,
Antiquarian Horological Society
(London, 1972), plus
Supplement, 1975.

Vaudrey Mercer, The Frodshams: the
Story of a Family of Chronometer
Makers, Antiquarian Horological
Society (London, 1981).

Charles Abdy, Ewell: a Surrey Village
that Became a Town, Surrey
Archaeological Society (2004).

Croydon Advertiser and Surrey County
Reporter, June 1935.

Post Office Magazine, August 1935.

Croydon Times, September–October
1935.

The Tatler, 9 October 1935.

Croydon Evening News,
November–December 1935.

Post Office Telephone Directory,
Croydon, 1935.

Watch and Clock Maker, April 1936.

William Pike, 'The GPO Speaking
Clock', Newnes Practical
Mechanics, May 1938.

The Guardian, 15 October 1996.

Chapter 9: 1939–1943

Guildhall Library Manuscripts
Section, London: Worshipful
Company of Clockmakers
papers, court minutes, 1941–4,
MS 2710/15.

Royal Greenwich Observatory
Archive, Cambridge: Harold
Spencer Jones papers, RGO
9/625, 'Papers on the Time
Service, 1943–1963'.

Royal Greenwich Observatory
Archive, Cambridge: Philbert
Melotte papers, RGO 74/6/2:
'Account of J. H. Belville and
GMT'.

South London Crematorium,
London: cremation register
and book of remembrance
('whole-page book'), p. 66.

W. A. Truelove & Son Ltd., Funeral
Directors, company records.

Will and grant, Elizabeth Ruth
Belville.

Death certificate, Elizabeth Ruth
Belville.

Personal communication, Bryan C.
Read, 2006 and 2008.

Personal communication, Francis
Ellison, 2008.

T. Lewes Sayer, Gog and Magog and I:
Some Recollections of 49 Years at
Guildhall, Sampson Low,
Marston & Co. Ltd. (London,
c. 1931).

A. J. Meadows, Greenwich Observatory,
vol. 2: Recent History, Taylor &
Francis (London, 1975).

Tony Jones, Splitting the Second: the
Story of Atomic Time, Institute
of Physics Publishing (Bristol,
2000).

Encyclopedia of Cremation, ed. Douglas
    Davies with Lewis Mates,
    Ashgate (Aldershot, 2005).
*The Longest Night: Voices from the
    London Blitz*, Gavin Mortimer,
    Phoenix (London, 2005).
South London Crematorium, company
    brochure, no date [c.2008].
Maidenhead Advertiser, 11 March
    1908.
*Third Supplement to The London
    Gazette of Friday the 26th of March
    1920*, 30 March 1920.
Evening News, 3 April 1929.
Astronomer Royal's Reports, Royal
    Observatory, Greenwich,
    1940–5.
Sutton Times and Cheam Mail, 17
    December 1943.
James Nye and David Rooney, 'Such
    Great Inventors as the Late Mr
    Lund: an Introduction to the
    Standard Time Company,
    1870–1970', *Antiquarian
    Horology*, December 2007.

PICTURE SOURCES AND CREDITS

Front endpaper
Underground Electric Railways of London. Map by Waterlow & Sons, 1907 (London Transport Museum © Transport for London).

Rear endpaper
The Port of London. Map from *Bacon's Large Scale Atlas of London and Suburbs, c.* 1912 (Guildhall Library © City of London.

Introduction
Evening News, 3 April 1929, p. 10

Chapter one
Punch, May 1883, p. 214 (D7065 © National Maritime Museum, London)

Chapter two
Private collection (© Graham Dolan)

Chapter three
Daily Graphic, 31 October 1892, p. 4 (PZ8786 © National Maritime Museum, London)

Chapter four
National Maritime Museum collection (B0718 © National Maritime Museum, London)

Chapter five
Fox Photos, in the *Daily Express,* 10 March 1908, p. 7 (© Getty Images)

Chapter six
Private collection (courtesy of Ian Henghes: ian@henghes.org)

Chapter seven
Private collection (Mary Evans Picture Library)

Chapter eight
Mitchell's Cigarettes Gallery of 1935, number 46, 1935 (Upacut Image Library)

Chapter nine
The Sketch, 17 July 1935, p. 146 (© 2008 The British Library)

ACKNOWLEDGEMENTS

So many people have helped, so many people to thank. Archivists and librarians across the south of England have let me in and led me to wonderful papers and pictures (in some cases, having to lead me by the hand). Very many people in this line of work helped me, and I thank them all. I will mention a few by name in particular (earnestly hoping that those not named will nevertheless know that I greatly appreciate what they did; I have listed elsewhere all the institutions I used). Adam Perkins, Godfrey Waller and colleagues at the Royal Greenwich Observatory archives in Cambridge University Library help me often and I always enjoy my trips to Cambridge. Arthur Holden and colleagues at Bromley Central Library have contributed greatly to my research into William Willett. Siân Wynn-Jones, Ray Martin, David Hay and colleagues at BT Archives have been brilliant on subjects including the speaking clock, the Standard Time Company and the Post Office. Anne Locker and Asha Marvin at the Institution of Engineering and Technology have been great friends too. Chris Bennett and colleagues at Croydon Central Library went out of their way to help me research Ethel Cain, and made me very welcome. The Science Museum Library staff, all of them, have always been real friends, and I have not forgotten their many kindnesses. The volunteers working at Ewell Public Library (whose names, I am ashamed to say, I forgot to ask, in my excitement at what they uncovered for me) were terrific. I can see why Ruth Belville liked Ewell so much.

Members of the families of people involved in this story have been unanimously supportive and generous with their time, memories and family papers. Bryan and Sheila Read, related to Maria Belville, welcomed me into their home and shared their stories and papers. Bryan met Ruth, back in the 1930s; it was an honour to shake his hand. Francis Ellison, another member of the family, helped enormously with genealogical research. George Willett and Ian Henghes (a grandson and a great-grandson respectively of William Willett) have always been extremely kind. Without the kindness of family members I would not have got nearly so close to the real people in this book.

Sir George White, keeper of the Clockmakers' Company collection, has been very supportive. He allowed the Belville pocket watch to be removed from its showcase and displayed at a lecture I gave at the London Guildhall in 2007. This meant a lot to me. Richard Stenning and Philip Whyte at Charles Frodsham & Co. Ltd. have offered advice from an early stage, and I have appreciated it. Richard let me have lots of information about Ruth's Charles Frodsham chronometer. David Parr helped me on railway clockmakers. Neil Johannessen, Alan Melia and Frank Milne advised me on aspects of telecommunications history. Mike Todd and his colleagues showed me round the BBC's Broadcasting House, a day I have never forgotten. David Bird hosted a visit to Rugby Radio Station, which blew my mind. Peter Whibberley, Sam Gresham and their many colleagues at the National Physical Laboratory have helped in more ways than I can even remember, let alone list. Tim Boon helped me look at William Willett in a different light (so to speak). Marjorie Hutchinson, who lives in Elliscombe Road, guided me to sources of Charlton history. David Boullin helped identify useful radio time references. Staff at Greenwich Council kindly tracked down Maria Belville's grave in Charlton Cemetery. Ian Ackhurst, churchwarden at Beddington, gave me valuable advice on how to find Ruth Belville's final resting place; the staff of W. A. Truelove, funeral directors, and those of the South London Crematorium, found what I was searching for. I visited both the cemetery and the crematorium and was made very welcome at both places. I am indebted to Betty Bartky and the late Ian Bartky, who helped me in ways too numerous to mention as I researched this story. I wish Ian could have seen the result, I really do.

At the National Maritime Museum, many colleagues including Gloria Clifton, Graham Dolan, Jonathan Betts, Nigel Rigby, Peter Linstead-Smith and Richard Dunn patiently supported me as I worked on this and related research projects. Peter is a volunteer at the Museum; he is also a retired submariner who is the Master of the Clockmakers' Company in 2009 and who invited me to speak about Ruth Belville at the Guildhall. Joan Mitchell and Peter Gosnell, our two other volunteers, have put up with me as I have wrestled with deadlines. I thank them, sincerely, for their patience, company and encouragement. Gloria, Graham and Jonathan also kindly read an early draft of this book, as did John Liffen at the Science Museum and James Nye of the Institute of Historical Research (also Secretary of the Electrical Horology Group). David Read, a radio time expert (amongst many other specialisms) read chapters too. To these people in particular I must express my most grateful thanks for all the help, encouragement and advice they have given me. My research project with James, a business historian, on the Standard Time Company, has been both stimulating and highly enjoyable. James has contributed to this book in many ways and I look forward to our next project together.

Lots of people contribute to the supportive working atmosphere of the Royal Observatory that inspires my passionate interest in its history: gallery assistants and guides, security and housekeeping staff, front of house and back-office colleagues, not to mention the astronomers, educators, object handlers, porters and technicians. It's a great team effort. Janet Small has kept me sane and sorted in numerous ways for which I thank her and Erik. In the Museum's publishing department, Emily Winter and Rachel Giles believed in this work and made it happen. I am grateful indeed; I hope I didn't disappoint. Abigail Ratcliffe, former project editor there, first suggested I turn my talks and papers into a book. Sara Ayad, picture researcher, did wonders securing the chapter-openers. Gary Steele kindly kept me supplied with inter-library loans of some highly esoteric articles. Pieter van der Merwe, the Museum's General Editor, performed a sensitive 'lift-and-dust' of my manuscript and used his exhaustive knowledge of Greenwich history to help in many ways. Bernard Dod shaped the manuscript still further and for this I am in his debt.

This research has been going on a while. Stephen Battersby at *New Scientist* revealed the story to that magazine's readers in 2006 and prompted a lot of encouraging interest. Editors Jeff Darken at *Antiquarian Horology*, John Hunter at *Clocks* and Polly Larner at *f@nmm* published my own articles. I value their faith in my writing. Everyone has tried their best to help me get this story about right, but the result is my responsibility: errors of fact, interpretation and emphasis are mine alone, of course, and I will be pleased to hear of them. I will also be delighted to hear from anyone connected to the story who might have more information to offer. In particular, I did not have time to track down the descendents of T. Lewes Sayer, because I only learned of his involvement at an advanced stage in the writing process. I'm also keen to learn more about Ethel Cain. Now, my memory is very fallible and I may have forgotten people who deserve to be thanked. If so, please forgive me and please let me buy you a drink.

Finally, and with love, I dedicate this book to my family, without whom I would be nothing. My mother, Pat; my father, Terry; my brother, Pete. The team. This is for you.

INDEX

*Page numbers in italics indicate
illustrations.*